Amos Emerson Dolbear

The Telephone

an Account of the
Phenomena of Electricity, Magnetism
and Sound
as Involved in Its Action

Elibron Classics
www.elibron.com

Elibron Classics series.

ISBN 1-4021-7087-4 (paperback)
ISBN 1-4021-1325-0 (hardcover)

This Elibron Classics Replica Edition is an unabridged facsimile of the edition published in 1877 by Lee & Shepard; Charles T. Dillingham, Boston; New York.

THE TELEPHONE:

AN ACCOUNT OF THE

Phenomena of Electricity, Magnetism, and Sound,

AS INVOLVED IN ITS ACTION.

WITH DIRECTIONS FOR MAKING

A SPEAKING TELEPHONE.

BY

PROF. A. E. DOLBEAR,
TUFTS COLLEGE,
AUTHOR OF "THE ART OF PROJECTING," ETC.

BOSTON:
LEE & SHEPARD, PUBLISHERS.
NEW YORK:
CHARLES T. DILLINGHAM.
1877.

FRANKLIN PRESS: RAND, AVERY, & CO.,
BOSTON.

PREFACE.

THE popular exhibitions of the speaking-telephone during the past six months, together with numerous newspaper articles, have created a widespread interest in the instrument; and it has been thought that a small book explanatory of its action would meet a public want.

It has seemed to be necessary to call attention to the various phenomena and inter-actions of the forces involved; and hence the author has attempted to make plain and intelligible the phenomena of electricity, magnetism, and sound. Cuts have been inserted where they could be useful in making the mechanical conditions more intelligible; and a table of tone-composition has been devised, which shows at a glance the constituents of the sounds of various musical instruments.

As the speaking-telephone, in which magneto-electric currents were utilized for the transmission of speech and other kinds of sounds, was invented by me, I have described at some length my first instrument, and have also given explicit directions for making a speaking-telephone which I know, by trial, to be as efficient as any hitherto made; but nothing in the book is to be taken as a dedication of the invention to the public, as steps have already been taken to secure letters-patent according to the laws of the United States.

A. E. DOLBEAR.

COLLEGE HILL, MASS.

THE TELEPHONE.

ELECTRICITY.

SOME of the phenomena of electricity are manifested upon so large a scale as to be thrust upon the attention of everybody. Thus lightning, which accompanies so many showers in warm weather in almost every latitude, has always excited in some individuals a superstitious awe, as being an exhibition of supernatural agency; and probably every one feels more or less dread of it during a thunder-shower, and this for the reason that it affects so many of the senses at the same time. The flash may be blinding to the eyes if near to us; the thunder may be deafening to the ears, and so powerful as to shake the foundations of the hills, and make the ground upon which we stand to sensibly move: these with the remembered destructive effects that have been witnessed, of buildings demolished and large trees torn to splinters in an instant, are quite sufficient to raise a feeling of dread in the strongest mind. In the polar regions, both north and south, where thunder-storms are less frequent, the atmospheric electricity assumes the form called the aurora borealis, or the aurora australis, according as it is seen north or south of the equator.

More than two thousand years ago it was noticed by the Greeks that a certain kind of a mineral which was thrown up on the shores of the Mediterranean Sea, when rubbed would attract light bodies, such as shreds of silk or linen and bits of paper. To this substance they gave the name of Elektron, and the property developed thus by friction was afterwards called electricity. In 1600 Dr. Gilbert, physician to Queen Elizabeth, published a book in which he described numerous experiments demonstrating that electricity could be developed by friction upon a great variety of substances, such as stones, gems, and resins. The first machine for developing electricity was made by Otto von Guericke of Magdeburg, about 1680.

His machine consisted of a ball of sulphur about six inches in diameter, which could be rotated. If the dry hand were held against the sulphur while it was being turned in a dark room, the sphere appeared to emit light: it also gave out a peculiar hissing or crackling sound. Newton experimented a little with electricity, and noticed that the rubber was an important element in developing electricity. He does not seem to have given to the subject the same attention that he gave to some other departments of science. Had he done so, it is probable that he would have advanced the study a hundred years; that is to say, he would probably have left it at the place where it actually was in 1790. So great were his abilities that in one lifetime he made greater additions to human knowledge than all the rest of mankind had made during the preceding thousand years. In the month of June, 1752, Franklin made that memorable experiment which immortalized him. He flew his kite to the thunder-cloud, practically asking the question of the lightning whether or not it was identical with electricity. The lightning came down the wetted twine to his hand, and proclaimed its identity.

For the next forty years the natural philosophers in both Europe and America only rung the changes upon what was known. They flew kites to the clouds; they made and charged Leyden jars, and discharged them through wires and chains and circuits of clasped hands, and studied the attractions and repulsions manifested by electrified bodies; but they added nothing of importance in the way of experiments.

In 1791 Galvani, a professor of anatomy at Bologna, announced a manifestation of electricity that was new and of a remarkable character, having its origin in the muscles of animals, and so was called animal electricity. He had some frogs' legs prepared for eating; by chance they were placed near an electrical machine with which Galvani was experimenting, so that a spark would occasionally pass to the legs, when they would contract as often as a spark passed to them. The motion was first observed by his wife, who called his attention to the phenomenon; and he very soon discovered that the thighs of a frog, skinned and suspended, made a very good electroscope. While experimenting in this way he made another and more important discovery; namely, that, when the muscles and nerves of the frog's leg were touched by pieces of two different metals, the leg would contract as before. Alexander Volta, another Italian professor, who had invented the electrophorus, and was possessed of great experimental skill, now turned his attention to the experiment of Galvani, and very soon discovered that the origin of the electricity that moved the frogs' legs was not in the legs themselves, but in the metals used. The first form of the galvanic battery was the result of Volta's investigations, and was called the Voltaic pile. This pile consisted of alternate disks of zinc, flannel, and copper, piled one on top of the other in constant succession in that order. The flannel was moistened with salt and water, or with diluted sulphuric acid. When the first zinc was connected with the last copper by means of a wire, a powerful current of electricity was obtained. This form of battery is not in use at all now, as much more efficient means are known for producing electricity; but this in 1800, when it was first made known in England, was very startling, and was one of those surprises which have been so frequent since then in the history of electricity.

Surprising things were done by Sir Humphry Davy, with a large Voltaic battery. Water was decomposed, and the metals potassium and sodium were first separated from their compounds with oxygen. Bonaparte had offered a prize of sixty thousand francs "to the person who by his experiments and discoveries should advance the knowledge of electricity and galvanism as much as Franklin and Volta did," and of "three thousand francs for the best experiments which should be made in each year on the galvanic fluid." This latter prize was awarded to Davy.

After Davy's successes in 1806, there was nothing of importance in an experimental way added to the knowledge of electricity, until 1820, when Oersted of Copenhagen announced that "the conducting wire of a Voltaic circuit acts upon a magnetic needle," and that the needle tends to set itself at right angles to the wire. This was a kind of action altogether unexpected. This observation was of the utmost importance; and at once the philosophers in Europe and

America set themselves to inquire into the new phenomenon. The laws of the motion of the magnetic needle when acted upon by a current of electricity traversing a wire were successfully investigated by M. Ampère of the French Academy. He observed that whenever a wire through which a current of electricity was passing was held over and parallel with a magnetic needle which was free to move, and therefore pointed to the north, if the current was moving *towards* the north, the north pole was deflected to the west; if the current was moving towards the south, the south pole of the magnet was deflected towards the west; and that in all cases the magnet tended to set itself at right angles to the current; also that this angular displacement depended upon the strength of the current. Thus originated the *galvanometer*, an instrument that not only detects the existence of an electric current, but enables us to determine its direction and its strength. Our present knowledge of electrical laws is due, in a very large measure, to observations made with this instrument. Of course it has been very much modified, and made almost incredibly sensitive: yet, in all galvanometers, the fundamental principle involved in their structure is that of the action of a current of electricity upon a magnet, which was first noticed by Oersted.

MAGNETS.

It is related by Nicander that among the shepherds who tended their flocks upon the sides of Mount Ida was one named Magnes, who noticed, that, while taking his herds to pasture, his shepherd's crook adhered to some of the rocks. From this man's name some have supposed the name *magnet* to have been derived. It is, however, generally believed to have received its name from the ancient city of Magnesia in Asia Minor, near which the loadstone or magnetic substance was found. This rock, which possesses the remarkable property of attracting and holding to itself small pieces of iron or steel, is now known to be one of the ores of iron, and is called magnetite by mineralogists. The iron is chemically combined with oxygen, and forms 72.5 per cent of its weight. There is another ore of iron, known as hematite, which contains seventy per cent of iron;

but the difference of two and a half per cent of iron in the ore is enough to make the difference between a magnetically inert substance, and one which may be able to lift a mass of iron equal to many times its own weight.

Sir Isaac Newton is said to have worn in a finger-ring a small loadstone weighing three grains, which would lift seven hundred and fifty grains, which is equal to two hundred and fifty times its own weight. The most powerful magnet now known is owned by M. Obelliane of Paris. It can lift forty times its own weight. Large pieces, however, do not support proportionally greater weights, seldom more than one or two times their own weight.

There are in many places in the world immense beds of magnetic iron-ore. Such are to be found in the Adirondack region in Northern New York, and in Chester County, Pennsylvania. The celebrated iron-mines of Sweden consist of it, and in Lapland there are several large mountains of it. It must not be inferred, that, because the mineral is called magnetite, all specimens possess the property called magnetism. The large masses seldom manifest any such force, any more than ordinary pieces of iron or steel manifest it: yet any of it will be attracted by a magnet in the same way as iron will be. The most powerful native magnets are found in Siberia, and in the Hartz, a range of mountains in Northern Germany.

When a piece of this magnetically endowed ore is placed in a mass of iron-filings, it will be seen that the filings adhere to it in greatest quantity upon two opposite ends or sides, and these are named the poles of the magnet. If the piece be suspended by a string so as to turn freely, it will invariably come to rest with the same pole turned towards the north; and this pole is therefore called the north pole of the magnet, and the action is called the directive action. This directive action was known to the Chinese more than three thousand years ago. In traversing those vast steppes of Tartary they employed magnetic cars, in which was the figure of a man, whose movable, outstretched arm always pointed to the south. Dr. Gilbert affirms that the compass was brought from China to

Italy in 1260, by a traveller named Paulus Venetus.

When a piece of hardened steel is rubbed upon a natural magnet, it acquires the same directive property; and, as the steel could be easily shaped into a convenient form for use, a steel needle has generally been used for the needle of a compass. The directive power of the magnet has been and still is of incalculable value to all civilized nations. Ocean navigation would be impossible without it, and territorial boundaries are fixed by means of it; but there are other properties and relations of a magnet, which have been discovered within the last fifty years, which are destined to be as important to mankind as that of the compass has been.

In 1825 William Sturgeon of Woolwich, Eng., discovered that if a copper wire were wound around a piece of soft iron, and a current of electricity sent through the wire, the soft iron would become a magnet, but would retain its magnetism no longer than while the current of electricity was passing through the coil. The magnetism developed in this way was called electro-magnetism, and the iron so wound was called an electro-magnet. The first electro-magnet was made by winding bare wire upon the soft iron. This method will not produce very strong magnets. In 1830 Prof. Henry insulated the wire by covering it with silk, and was the first to produce powerful magnets.

On a soft iron bar of fifty-nine pounds weight he used twenty-six coils of wire, thirteen on each leg, all joined to a common conductor by their opposite ends, and having an aggregate length of seven hundred and twenty-eight feet. This apparatus was found able to sustain a weight of twenty-five hundred pounds. This electro-magnet is now owned by Yale College.

The power of the electro-magnet is enormously greater than that of any permanent magnet. A permanent magnet made by Jamin of Paris, which is made up of many strips of thin steel bound together, and weighing four pounds, is able to support a weight of one hundred pounds; but Dr. Joule made an electro-magnet, by arranging the coils to advantage, that would support thirty-five hundred times its own weight, or one hundred and forty times the proportionate load of Sir Isaac Newton's ring magnet.

THE GALVANIC BATTERY.

The original form of the galvanic battery as devised by Volta, and modified but little during thirty years, consisted of a cell to contain a fluid, which was usually dilute sulphuric acid, in which two plates of different metals were immersed: the metals used were generally plates of zinc and copper, or zinc and silver. Such plates, when first placed in the liquid, will give a very good current of electricity; but it will not last long. The reason of this is easy to understand. Whenever a current of electricity is generated by chemical action of a liquid upon two different metals, there is always some decomposition of the liquid, and this decomposition takes place upon the plates themselves; and the liberated gases *adhere to the plates, and prevent further contact with the acid;* at the same time, the gases themselves act upon the plates, and generate a current of electricity in the opposite direction. This will of course interfere with the first current; and very soon the battery is useless until the plates have been withdrawn from the liquid. This physico-chemical process that takes place in such a battery is called the *polarization of the plates.*

The accompanying figure will help one to understand the actions going on in a battery cell of the kind mentioned. Let Pt represent a plate of platinum, and Zn a plate of zinc, both placed in a vessel containing hydrochloric acid, which is also represented by the symbols HCl.

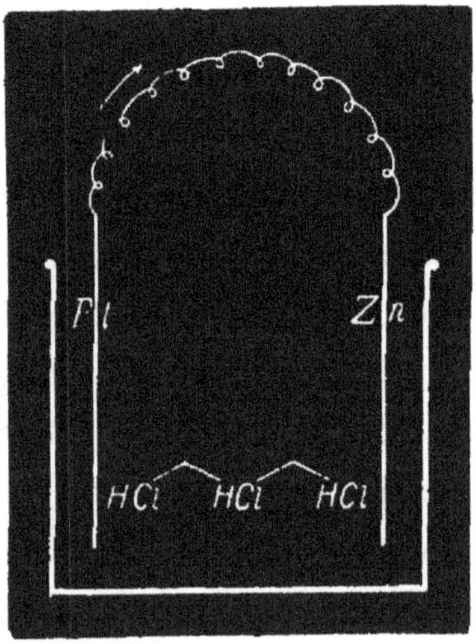

FIG. 1.

As such molecules are extremely minute, there will of course be an immense number of them between the plates. The plates are now to be connected by a wire running between them through the air. As soon as these conditions are fulfilled, a hissing sound will be heard coming from the cell, and bubbles of gas will be seen to rise from the platinum plate: these bubbles prove upon analysis to be bubbles of hydrogen. At the same time the zinc will begin to dissolve, forming what proves by analysis to be the chloride of zinc; and at the same time a current of electricity travels through the wire from the platinum to the zinc. The quantity of electricity that is thus generated is strictly proportionate to the quantity of hydrogen liberated, which is also proportionate to the weight of zinc dissolved; and this, in turn, is proportionate to the surface of the metals exposed to the action of the acid. Now, it happens under such circumstances as the above, that the liberated hydrogen adheres very strongly to the platinum, as there is nothing for it to unite with chemically; and therefore the plate will very soon be visibly covered with bubbles, which may be scraped off with a feather or a swab, but only to have the same thing repeated.

This coating of bubbles will prevent the acid from touching the plate, and so practically diminishes the surface of it; but the quantity of electricity generated being proportionate to the surface exposed to the chemical action, it will be understood at once how such polarization of the plates must soon bring the battery to a standstill.

In 1836 Prof. J. F. Daniell of London contrived a battery, which has been called the Daniell Cell, in which the metal (copper) that had the hydrogen liberated upon it was separated by a porous cell from the zinc. The zinc was immersed in dilute sulphuric acid, and the copper in an acid solution of blue vitriol (copper sulphate). The porous cup did not prevent the electricity from passing, nor the decomposition from taking place; but the hydrogen, which in this case would have been liberated at the copper plate, at once united with oxygen there, which it got by decomposing the copper sulphate: hence water was formed, and copper was deposited upon the copper plate; and, being an excellent conductor, the battery would keep up a strong action for a long time.

Mr. Grove, also of London, in 1839 invented a battery which still goes by his name, in which the hydrogen plate is of platinum immersed in strong nitric acid, enclosed also in a porous earthen cell; and this, in turn, is plunged into a vessel containing dilute sulphuric acid and the zinc. In this case the liberated hydrogen immediately decomposes the nitric acid, which readily parts with its oxygen; water is the product, as in the other case, and the nitric acid loses strength. Strips of carbon have been substituted for the platinum, and this is called the Bunsen battery. It is otherwise like the Grove battery; it gives a very powerful and constant current: and it is by the use of one or the other of these batteries, that most of the experiments in electricity are performed in institutions of learning, and, until lately, most in use for telegraphic purposes.

OTHER MEANS FOR GENERATING ELECTRICITY.

THERMO-ELECTRICITY.

If two strips of different metals, such as silver and iron, be soldered together at one end, and the other ends be connected with a galvanometer, on heating the soldered junction of the metals it will be found that a current of electricity traverses the circuit from the iron to the silver. If other metals be used, having the same size, and the same degree of heat be applied, the current of electricity thus generated will give a greater or a less deflection, which will be constant for the metals employed. The two metals generally employed are bismuth and antimony, in bars about an inch long and an eighth of an inch square. These are soldered together in series so as to present for faces the ends of the bars, and these often number as many as fifty pairs. Such a series is called a thermo-pile. This method of generating electricity was discovered by Seebeck of Berlin in 1821, but the thermo-pile so much in use now in heat investigations was invented by Nobili in 1835. The strength of this current is not very great, a single Daniell cell being equal to nine pairs of the strongest combination yet discovered, namely, the artificial sulphuret of copper with German silver.

MAGNETO-ELECTRICITY.

It has already been mentioned, that Oersted found that a magnet when free to turn tended to set itself at right angles to a wire in which a current of electricity was passing, thus demonstrating some inter-action between electricity and magnetism; but it remained for Faraday to discover the converse fact, namely, that a magnet moving across a wire, the ends of which were connected with a galvanometer or otherwise closed, originated a current of electricity in the wire, the direction of which depended upon the direction of the movement of the magnet. If the wire was coiled into a hollow helix, the magnet in moving through the helix moved across, that is, at right angles to all the turns of the helix; and each complete turn added to the intensity of the current. This will be understood by reference to the diagram, Fig. 2. Let G be a galvanometer connected with the wires from a helix; N S, a permanent bar magnet. If the magnet be thrust into the coil, a current of electricity will traverse the helix, wire, and galvanometer, and the needle will indicate its direction. If the magnet be now withdrawn, a current will move in the opposite direction through the whole circuit. The electricity that is thus originated is said to be induced. The quantity of electricity that can be induced thus is almost unlimited, depending upon the size and strength of the magnet, the size of the wire, and the length of wire in the coil. There are now many forms of machines for developing electricity from the motion of coils of wire in front of the poles of

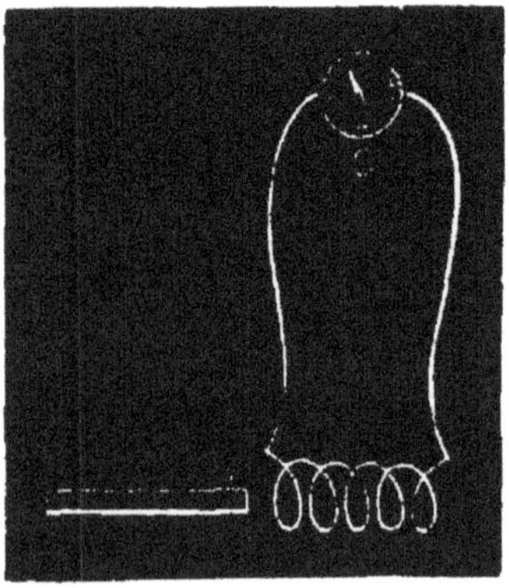

FIG. 2.

permanent magnets. They are generally called magneto-electric machines. The action involved in these machines is so important in its bearing upon telephony as to necessitate a fuller description of them.

MAGNETIC INDUCTION.

Let N S, Fig. 3, be a bar of hardened steel rendered permanently magnetic. If now there be

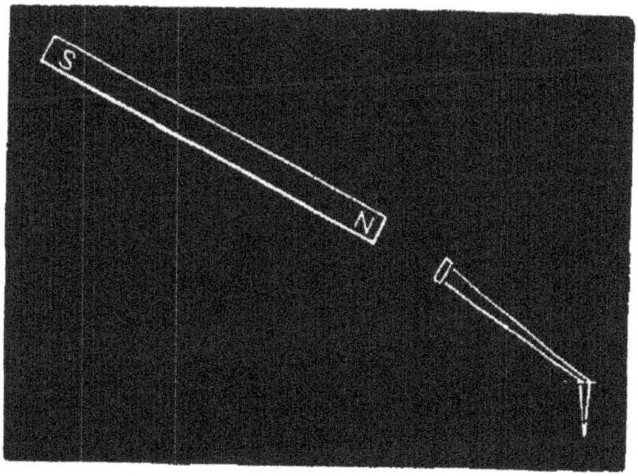

FIG. 3.

brought near to it a board-nail, the latter will become a magnet through the *inductive* action of the first magnet. This induced magnetism may be demonstrated by bringing a tack or other bit of iron to the end that is farthest from the permanent magnet; the tack will adhere to the nail, but will fall off when the nail is removed from the neighborhood of the magnet. By testing the polarity of the nail, it will be found that the end nearest the magnet will be a south pole if the magnet has its north pole towards it, in all cases having a polarity opposite to that of the pole acting upon it. The strength of this induced magnetism thus developed depends upon the distance apart of the magnet and the iron, being at its maximum when the two touch. But the tack itself is also made a magnet, and will attract another tack, and that one still another, the number which can be thus supported being dependent upon the strength of the first or inducing magnet.

Suppose now that we should wind a few feet of wire about the nail, and fasten the two ends of the wire to an ordinary galvanometer, and then make the nail to approach the permanent magnet. The galvanometer needle would be seen to move as the nail approached; and, if the latter were allowed to touch the magnet, the movement of the needle would suddenly be much hastened, but would directly come to rest, showing that, so long as there is no motion of the nail towards or away from the magnet, no electricity is moving in the wire, although the nail is a strong magnet while it is in contact with the permanent magnet. If the nail be now withdrawn, the two phenomena happen as before: that is to say, as the nail recedes it loses its magnetism; and the giving-up of its magnetism induces a current of electricity through the wire in the opposite direction to that it had when the nail approached. The current of electricity in the opposite direction is indicated by the galvanometer needle, which moves according to Ampère's law mentioned on a preceding page.

It may be noted here that we have an effect quite analogous to that already mentioned on page 21 as the experiment of Faraday. In one case a permanent magnet is thrust into a coil of wire, and in the other a piece of iron is made a magnet while enclosed in a coil. In each case there is generated a current of electricity *which lasts no longer than the mechanical motion of the parts lasts.*

MAGNETO-ELECTRIC MACHINES.

Such transient currents are practically useless, and several devices have been invented to make the flow continuous. The common form of machine for doing this may be understood by reference to the diagram.

N S, Fig. 4, is the permanent magnet, which is

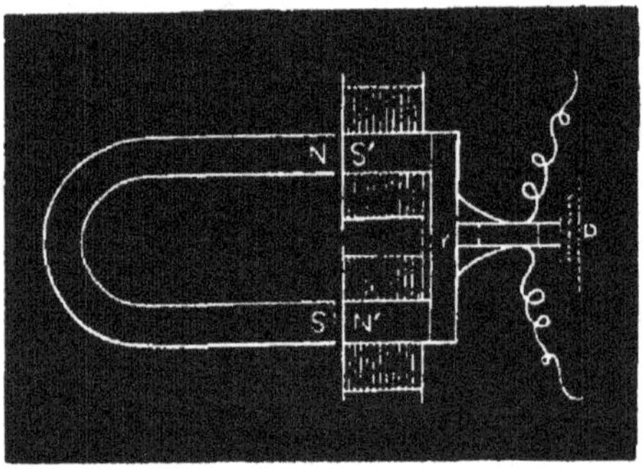

FIG. 4.

bent into a U form in order to utilize both poles. N′ and S′ are short rods of soft iron fastened into a yoke-piece Y, also of soft iron. Coils of wire surround each of the rods as represented, the ends of the wires connecting with each other and with what is called a pole-changer. The whole of this part is capable of revolving upon an axis P Y by a pulley at P. The action is as follows: From their position, the soft-iron rods N′ S′ must be magnets through the inductive action of the permanent magnet, just as the nail was made a magnet in like position. So long as the parts have the relative position shown in the figure, and there is no motion, no electricity can be developed; but, if the axis P Y be turned, S′, which represents the polarity of the rod opposite N, will be losing its induced magnetism; and, when half a revolution has been made, that same pole will be where N′ now is; but it will then have N′ polarity instead of S′; that is, it has been losing south polarity as it receded from N, and gaining north polarity as it approached S: hence a current of electricity has steadily been flowing through the coil in one direction. At the same time, the other rod N′ has passed through similar phases; and its enveloping coil has had a current of electricity induced in it in the same direction as in the first coil. This doubles the intensity of the current; and the whole is conducted by the connecting-wires where the current is wanted. Machines have been built upon this plan, that contained fifty or sixty powerful compound permanent magnets, and as many wire coils, needing a steam-engine of eight or ten horse-power to run them.

A less cumbersome and much more efficient magneto-electric machine has been made by changing the form of the soft iron armature to something like a shuttle, and winding the wire inside of it. This is called the "Siemen's Armature." The latest pattern of such machines is known as the *Gramme;* and its peculiarity consists in the substitution of a broad ring of soft iron for the armature. About this ring a good many coils, of equal lengths, of insulated copper wire are wound in such a manner that one-half of any turn in the wire goes through the inside of the ring, making the coils longitudinal. The whole of the armature thus prepared is fixed upon a shaft, so as to permit rotation, and fixed between the poles of a powerful Jamin magnet. The ends of the coils are connected with conductors upon the axis; and, when the armature thus constructed is rotated, a very constant and powerful current of electricity flows in a single direction, unlike the other forms. It is stated, that, with one-horse power, a light can be obtained equal to that from a battery of fifty Grove cells.

SECONDARY CURRENTS.

So long ago as 1836 it was noticed by Prof. Page of Salem, that, whenever a current of electricity was made to flow in a coil of wire, another current in the opposite direction was induced in a coil that was parallel with the first; and also, when the current in the first was broken, another current in the second coil would flow in the opposite direction to the former one. These currents, which are called secondary currents, are very transient. No current at all flows save at the instant of making or breaking the current. In this respect, we are reminded of the behavior of the soft iron within the coil, which gives origin to a current of electricity when it is made to approach a magnet or recede from it, but gives no current so long as it is still.

These secondary currents were investigated by Prof. Henry, resulting in the discovery of many curious and interesting phenomena. It will be sufficient here for me to refer to what are called induction coils, which are developments of the principles involved in electro-magnetism and electro-induction. Imagine a rod of soft iron of any size to be wound with a coil of wire, the ends of the wire to be so left that they may be connected with a galvanic battery. Around this coil let another coil be wound of very fine and well-insulated wire; the terminal wires of it to be left adjustable to any distance from each other. Now, upon making connection with a battery to the primary coil, there will be two results produced simultaneously. First, the soft iron will be rendered magnetic; and, second, a current of electricity will be generated in the secondary coil; and the strength of this secondary current is very much increased by the inductive action of the soft iron that has been made a magnet. When the battery current is broken, the iron loses its magnetism, and a current of electricity is again started in the secondary coil in the opposite direction. The energy of this derived current is so great that it will jump some distance through the air, and thus is apparently unlike the electricity that originates in a battery.

An induction coil made by Mr. Ritchie for the Stevens Institute at Hoboken, N.J., has a primary coil of 195 feet of No. 6 wire. The secondary coil is over fifty miles in length, and is made of No. 36 wire, which is but .005 of an inch in diameter. This instrument has given a spark twenty-one inches in length, with three large cells of a bichromate battery.

Mr. Spottiswood of London has just had completed for him the largest induction coil ever made. It has two primary coils, one containing sixty-seven pounds of wire, and the other eighty-four pounds, the wire being .096 inch in diameter. The secondary coil is two hundred and eighty miles long, and has 381,850 turns. This coil is made in three parts, the diameter of the wire in the first part being .0095 inch; of the second part, .015; and the third part, .011. With five Grove cells this induction coil has given a spark forty-two inches long, and has perforated glass three inches thick.

The electricity thus developed in secondary coils is of the same character as that developed by friction; and all of the experiments usually performed with the latter may be repeated with the former, many of them being greatly heightened in beauty and interest. Such, for instance, are the discharges in vacuo in Geisler tubes, exhibiting stratifications, fluorescence, phosphorescence, the production of ozone in great quantity, decomposition of chemical compounds, &c.

The electricity developed by friction upon glass, wax, resin, and other so-called non-conductors, has heretofore been called static electricity, for the reason that when it was once originated upon a surface it would remain upon it for an indefinite time, or until some conducting body touched it, and thus gave it a way of escape. Thus, a cake of wax if rubbed with a piece of flannel, or struck with a cat-skin or a fox-tail becomes highly electrified, and in a dry atmosphere will remain so for months. Common air has, however, always a notable quantity of moisture in it; and, as water is a conductor of electricity, such damp air moving over the electrified surface will carry off very soon all the electricity.

Again, the electricity developed through chemical action in a battery and through the inter-action of magnets and coils of wire has been called dynamic electricity, inasmuch as it never appeared to exist save when it was in motion in a completed circuit. This, however, is not true; for if one of the wires from a galvanic battery be connected with the earth, and the other wire be attached to a delicate electrometer, it will be found that the latter gives evidence of electrical excitement in the same manner as it does for the electricity developed by friction in another body. This is sometimes called *tension*, and is very slight for a single cell; but in a series of cells it becomes noticeable in other ways. Thus when the terminals of a single cell are taken in the hands, no effect is perceived: if, however, the terminals of a battery consisting of forty or fifty cells be thus taken, a decided shock is felt, not to be compared though with the shock that would be felt from the discharge of a very small Leyden jar. The shock from several hundred cells would be very dangerous.

It was formerly doubted that the electricity would pass between the terminals of a battery without actual contact of the terminals. Gassiot first showed that the spark would jump between the wires of a battery of a large number of cells before actual contact was made. Latterly Mr. De La Rue has been measuring the distance across which the spark would jump, using a battery of a large number of cells.

I give his table as taken from the "Proceedings of the Royal Society:"—

Cells.	Striking distance.
600	.0033 inch.
1,200	.0130 "
1,800	.0345 "
2,400	.0535 "

This table shows that the striking distance is very nearly as the square of the number of cells. Thus, with 600 cells the spark jumped .0033 inch; and with double the number of cells, 1,200, the spark jumped .0130 inch, or within .0002 of an inch as far as four times the first distance.

This leads one to ask how big a battery would be needed to give a spark of any given length, say like a flash of lightning. One cell would give a spark .00000001 inch long, and a hundred thousand would give a spark 92 inches long. A million cells would give a spark 764 feet long, a veritable flash of lightning. It is hardly probable that so many as a million cells will ever be made into one connected battery, but it is not improbable that a hundred thousand cells may be. De La Rue has since completed 8,040 cells, and finds that the striking distance of that number is 0.345 inch, a little more than one-third of an inch. He also states that the striking distance increases faster than the above indicated ratio, as determined by experimenting with a still larger number of cells.

These experiments and many others show that there is no essential difference between the so-called static and dynamic electricity. In the one case it is developed upon a surface which has such a molecular character that it cannot be conducted away, every surface molecule being practically a little battery cell with one terminal free in the air, so that when a proper conductor approaches the surface it receives the electricity from millions of cells, and therefore becomes strongly electrified so that a spark may at once be drawn from it.

WHAT IS ELECTRICITY?

THEORIES.

Numerous attempts have been made to explain the phenomena of electricity. As a general thing, these phenomena are so utterly unlike other phenomena that have been explained and are easily intelligible, that it has quite generally been taken for granted, until lately, that something very different from ordinary matter and the laws of forces applicable to it must be involved in the phenomena themselves. Consequently the term *imponderable* was applied to it, — something that was matter minus some of the essentials of matter; and as it was apparent that, whatever it was, it moved, apparently flowed, from one place to another, the term *fluid* was applied to it, a term descriptive of a certain form of matter. Imponderable fluid was the descriptive name applied to electricity. Newton supposed that an excited body emitted such a fluid that could penetrate glass. When the two facts of electrical attraction and repulsion had to be accounted for, two theories were propounded,— one by Benjamin Franklin, the other by Dufay. Franklin supposed that electricity was a subtle, imponderable fluid, of which all bodies contained a certain normal quantity. By friction or otherwise this normal quantity was disturbed. If a body received more than its due share, it was said to be positively electrified: if it had less than its normal quantity, it was said to be negatively electrified. Franklin supposed this electric fluid to be highly self-repulsive, and that it powerfully attracted the particles of matter.

According to Dufay, there are two electric fluids, opposite in tendency but equal in amount. When associated together in equal quantities, they neutralize each other completely. A portion of this neutral compound fluid pervades all matter in its unexcited state. By friction or otherwise this compound fluid is decomposed, the rubber and the body rubbed exchanging equal quantities of opposite kinds with each other, leaving one of them positively, the other negatively electrified. These two fluids were supposed to be self-repulsive, but to attract each other: so that, if two bodies be charged with either positive or negative electricity, such bodies would mutually repel each other; but if one was charged with positive, while the other was charged with negative electricity, the two bodies would mutually attract each other.

Either of these two theories may be used to illustrate the phenomena, and so have done good service in systematizing the facts. It is evident that both of them cannot be true, and it is in the highest degree probable that neither of them is true.

Some have supposed that there was a kind of electric atmosphere about every atom of matter; and still another theory, now advocated by Edlund of Stockholm, assumes that electricity is identical with the ether by which radiant energy, light and heat, is transmitted.

Before a correct judgment can be formed of the nature of any force, it is necessary to know what it can do, what kind of phenomena it can produce. Let us, then, take a brief survey of what electricity can do.

1st, It can directly produce *motion*, through the attractions and repulsions of electrified bodies, — as indicated by electrometers, the rotation of the fly-wheel, the deflection of the galvanometer needle. It has been proved by the mathematical labors of Clausius, and confirmed by experiment, that, when electricity performs any mechanical work, so much electricity is lost, annihilated as electricity.

2d, It can directly produce *heat*, as shown by passing a sufficient quantity of electricity through a fine platinum wire: the wire becomes heated, and glows, and it may even be fused by the intensity of the heat. The heat developed in the so-called electric arc is so great as to fuse the most refractory substances. If a current of electricity from a battery be sent through a thermo-pile, one of the faces of the pile will be heated. The heat of the spark from a Leyden jar may be made to ignite gunpowder, and dissipate gold into vapor. The heat produced by lightning is seen when a live tree is struck by a powerful flash: the sap of the tree is instantly converted into steam of so high a tension as to explode the tree, scattering it in small fragments over a wide area. The tips of lightning-rods often exhibit this heating effect, being fused by the passage of too great a quantity of electricity.

In the early part of the present century it was demonstrated by Count Rumford, and also by Sir Humphry Davy, that heat was but a form of molecular motion. Since then the exact relations between the motion of a mass of matter and the equivalent heat have been experimentally determined by Joule, so that the unit of heat may be expressed in the motion of a mass of matter. This is deducible from a more general law, known as the conservation of energy. The application in this place is, that whenever heat appears through electric action, as in the above-mentioned places, we know that it still is only *motion* that is the product, only that this motion is now among the molecules of the body, instead of the motion of the whole body in space, as when a pith-ball moves, or a galvanometer-needle turns.

3d, It can directly produce *light*. This is seen in every spark from an electric machine, in the flash of lightning, and in the electric light.

It has been shown in numberless ways, that there is no essential difference between light and heat, and that what we call light is only the active relation which certain rays of radiant energy have to the eyes. In order to make this plain, suppose that a beam of light, say from the sun, be permitted to fall upon a triangular prism of glass: at once it is seen that the beam is deflected, and instead of appearing a spot of white light, as it did before it was deflected, it now appears as a brilliant band of colors, which is called the solar spectrum. If now this spectrum be examined as to the distribution of heat, by moving a thermo-pile through it from the blue end towards the red end, it will be noticed that the galvanometer-needle will be but slightly deflected at the blue end; but, as the thermo-pile is moved, the deflections are greater until it is past the red end, where the heat is greatest. On this account it has been customary to say that the red end of the spectrum was the heating end. With various pieces of mechanism the rays may be separated from each other, and measured; and then it appears that a red ray of light has a wave length of about $\frac{1}{37000}$ in., and the violet ray about $\frac{1}{60000}$ in. The rays beyond the red have also been measured, and found to be greater in length uniformly as one recedes from the visible part of the spectrum.

In like manner, beyond the blue end the wave lengths become shorter and shorter; and in each of these directions the spectrum that is invisible is much longer than the visible one. Now, it has also been found that where a prism of glass or other material is used to produce a spectrum, it distributes the rays very unevenly; that is, towards the red end of the spectrum they are very much crowded, while towards the blue end they are more dispersed. Hence, if one were measuring the heating power of such a spectrum, many more rays would fall upon an equal surface of the thermo-pile at the red end than at the blue end; therefore the indications of the galvanometer would be fallacious. Before any thing definite could be known about the matter, it would plainly be necessary to work with an equal dispersion of all the rays. This was effected a few years ago by Dr. Draper of New

York. He took the spectrum produced by diffraction instead of refraction, and measured that. In that way it was found that the heating power of the spectrum is equal in every part of it; and hence the pictures in treatises on physics that represent the heating power of the spectrum to be concentrated at the red end is not true save where the spectrum is irregularly produced. As for vision, the mechanical structure of the eye is such that radiant vibrations having a wave length between $\frac{1}{37000}$ in. and $\frac{1}{60000}$ in. can affect it, while longer or shorter wave lengths can not. Such waves we call light, but it is not at all improbable that some animals and insects have eyes adapted to either longer or shorter wave-lengths; in which case, what would be perfectly dark to us would be light to them. It is a familiar enough fact, that many animals, such as dogs, cats, rats, and mice, can see in the night. Some horses may be trusted to keep in the road in a dark night, when the driver cannot see even the horse itself. This has usually been accounted for by saying that their eyes are constructed so as to collect a greater number of luminous rays.

It is much better explained by supposing their eyes to be constructed to respond to wave-lengths either greater or less than those of mankind.

A ray of light, then, consists of a single line of undulations of a definite wave length, such that if it falls upon the eye it will produce sight; if it falls upon a thermo-pile it heats it by just the same quantity that another wave-length would heat it; if it falls upon matter in unstable chemical relations, it will do chemical work, depending upon the kinds of matter. A red ray is as effective for some substances as a violet ray is for others. The statement, then, so often lately made to do certain analogical work, namely, that a ray of light consists of three distinct parts, which may be separated from each other, and are called heat, light, and chemical properties, is simply untrue. What a ray will do, depends upon what kind of a structure it falls on; and when it has done that work, of whatever kind it may be, it ceases to exist as a ray.

If, therefore, electricity can directly produce light, it is simply producing *motion*, as in the case of heat, the motion being of such a sort that the eyes of men are affected by it.

4th, It can produce *magnetism.* A current of electricity passing through a coil of wire makes such a coil a magnet, which will set itself in the direction of the magnetic meridian of the earth; and, if a bar of soft iron be placed in the coil, it becomes the familiar electro-magnet; and, if hardened steel be put in it, it becomes a permanent magnet.

This leads to the inquiry as to what magnetism is. We know that it can produce motion by its moving at a distance a piece of iron or another magnet. It will also sustain a mass of matter against gravity or some other contrary force. Through such mechanism as magneto-electric machines it produces electricity in great abundance, which again can be used to produce any of the effects of electricity, — moving bodies by attraction or repulsion, generating heat or light, or again making a magnet. But as all of these are but varied forms of motion, either of a mass as a whole, or molecular, can it be doubted for an instant, that what we call magnetism is but some form of motion? Must it not be either some form of matter, or some form of motion? If it were a form of matter, then a magnet would only be permanent so long as it was not used; for use implies consumption of the force; and, if this be matter in any form, then in a given mass of matter there can be but a definite quantity of such magnetic matter, and consumption must lessen that quantity. As a matter of fact, there is no perceptible lessening of the power of a magnet when it is properly used. It is also a matter of fact, that neither motion of a mass, nor electrical effects, nor any other, can be produced by the action of a magnet alone. It is only when some form of motion has been added to its own property, that we get any kind of an effect from it: hence all effects due to its action are *resultants* of two forces, one of them being common motion of a mass of matter, and the other the energy of the magnet. Hence we infer that a magnet is a mechanism of such a structure as to change the direction and character of the motion which acts upon it. When the wheel of a common electrical machine is turned, the product is electricity,—a force very different from that which originates it. Ordinary mechanical motion *goes in;* electricity *comes out*, the latter being a modified motion due to the physical structure of the machine. In like manner, a magnet may be considered as a machine by means of which mechanical motion may be converted into some other form of motion. It is evident that molecular structure is chiefly concerned in this. If a bar of iron that exhibits no evidence of magnetism whatever be subjected to torsion, it will immediately become a magnet with poles dependent upon the direction of the twist. This developed magnetism will re-act upon a coil of wire, and so move a galvanometer needle. If the bar be permitted to recover its original condition, it will lose its magnetism, which will at once re-appear upon twisting the rod again. Now, when the rod is twisted, it is evident that there is a molecular strain in certain directions throughout the mass. The converse experiment illustrates the same thing. It has been found, that when a rod of iron is made magnetic by the action of a current of electricity circulating about it, and at the same time passing longitudinally through it, the rod is slightly lengthened and twisted in a direction that depends upon the direction of the current. Moreover, if a permanent magnet be heated to a red heat, its magnetism is destroyed; for such a heat allows the molecules to freely arrange themselves without any external constraint. Also, if a permanent magnet be suspended so as to give out a musical sound when it is struck, the magnetism will be much weakened by making it thus to vibrate. In this case, as in the other, the vibrations affect every molecule, and so enable them to re-adjust themselves to the positions they held before being magnetized. The same thing happens when a bar of iron is made magnetic through the inductive action of the earth. When this bar is held in the direction of the magnetic dip, it becomes but very slightly magnetized; but, if it be so held that when it is struck with a hammer it will ring, that is, give out a musical sound, it will at once become decidedly magnetic. Evidently the earth's action tends to set the molecules of the mass in a new position, but cohesion prevents them from assuming it. When the molecules are made to vibrate, they can assume such new positions more readily. The molecules of a magnet, then, are differently arranged from those in an unmagnetized piece of iron or steel; and, for every new arrangement of the molecules of a mass of any kind, we always have some new physical property developed. The same identical substance may appear as charcoal, coke, plumbago, anthracite coal, and diamond. Hence a magnet is a machine in which other forces acting upon it are transformed in character, and re-appear as attractions and repulsions of other kinds of matter: this transformation cannot take place, and hence magnetism cannot become apparent, only upon the condition of another force acting in concert with it; and, if at any time it may seem to be acting without such external force, it is done at the expense of the heat it has absorbed, and therefore the magnet must at such time be losing temperature proportional to the work done. This I have discovered to be true by making a magnet to exert its force in front of a thermo-pile, which uniformly exhibits a cooled face under such conditions. What the particular form of the motion may be that we call magnetism, is not yet made out; but that it is some form of motion, is very evident. The following experiments may throw some light upon it. Last August Mr. Kerr read a paper before the British Association of Science, in which was detailed the following experiment: The pole of an electro-magnet was nicely polished so as to reflect light like a mirror. A beam of sunlight was permitted to fall upon it, and be reflected to a convenient place for examination. A current of electricity was sent through the coil, which of course rendered the iron magnetic; and it was noticed that the light that was reflected from the pole was circularly polarized: that is, the motion of a ray, instead of being a simple undulatory movement, was now made to assume such a motion as the water from a garden-hose has when the nozzle is swung round in a circle while the water is escaping from it. After reading the account of it, it occurred to me that the converse experiment might be tried; that is to say, the effect of a circularly polarized beam of light upon a piece of steel. By concentrating a large beam of ordinary plane polarized light with a quartz lens, and passing it through a quarter wave-plate at the proper angle, a powerful beam of circularly polarized light was obtained. At the focus of this beam a fine cambric needle without magnetism was placed so that the light passed it longitudinally. Ten minutes' exposure was sufficient to make it decidedly magnetic. Hence I infer that the motions which we call magnetic attractions and repulsions may be quite analogous to such helical motions; also, that these motions exist in ether, and evidently may be either right-handed or left-handed. Wind up on a pencil a piece of wire twelve or fifteen inches long, making a loose spiral. Bring the two ends of the spiral together; and note first that one is twisted to the right, the other to the left. If they be twisted into each other, they will advance very easily; but if a right-handed spiral were to be interlocked with another like it, and both turned in the direction of their spiral, they would separate rapidly. Applying this conception to a magnet, we might suppose that such spiral motions will be set up in the ether by the magnet, and that such motions re-acting upon ordinary matter affect it as attraction and repulsion; and thus we should have at least a conceivable mechanical explanation of the phenomenon.

There are numberless experiments which might be given to further exhibit the relation of mass motion to magnetism, but a single one more must

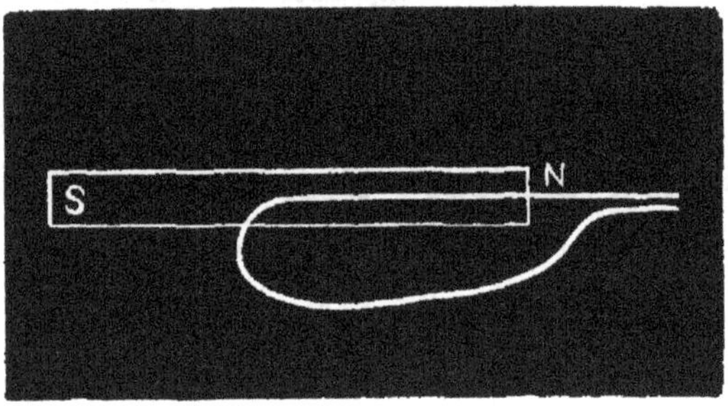

FIG. 5.

suffice. No rotation of a magnet upon its own axis can produce any effects upon a current that is exterior to it; but if a loop of wire be kept stationary adjacent to a magnet, as in Fig. 5, while the magnet revolves, a current of electricity is produced; and if the magnet be kept stationary, and the loop revolves, a current will also be produced, but in the opposite direction. Here, as in all the other cases, no electricity is originated, save when motion is imparted to one or other of the parts. This experiment is due to Faraday.

From all these cases we can come to but one conclusion, that both electricity and magnetism are but forms of motion; electricity being a form of motion in ordinary matter, for it cannot be made to pass through a vacuum, while magnetism must be a form of motion induced in the ether, for it is as effective in a vacuum as out of it; electricity always needing some material conductor, magnetism needing no more than do radiant heat and light.

VELOCITY.

Measurements have been made of the velocity of electricity; both that of high tension, such as the spark from a Leyden jar, and also that from a battery. The former was found to have a velocity over 200,000 miles a second, while the electricity from a battery may move as slowly as 15,000 or 20,000 miles a second; but this is very largely a matter of conductors. Its velocity is seldom above 30,000 miles a second on ordinary telegraphic lines. If the electricity be used to give signals, as in ordinary telegraphy, the time required varies nearly as the length of the line, and in any case is a much greater quantity. Prescott in his work on the telegraph states that "the time required to produce a signal on the electro-magnet at the extremity of a line of 300 miles of No. 8 iron wire is about .01 seconds, and that this time increases in a greater proportion than the length of the line; for example, on a line 600 miles in length it amounts to about .03 seconds." He also states that it varies much with the kind of magnet used, some forms of magnets being much more sensitive than others for this work.

Wheatstone proved a good many years ago that the duration of the electric spark was less than one millionth of a second. When a swiftly moving body can only be seen by an electric spark, or flash of lightning, it looks as if it were quiescent. Thus a train of cars rushing along at the rate of forty or fifty miles per hour appears sharply defined, — even the driving-wheels of the locomotive can be seen in detail, which is impossible in continuous light, — and all seems to be standing still. In like manner will the sails of a windmill, which may be turning at a rapid rate, be seen apparently at rest. This is because in the short time during which they are illuminated they do not appreciably move.

I am not aware that any attempt has been made to measure the velocity of magnetism. If, however, it be a form of motion in ether, it is probable that the velocity is comparable to the velocity of radiant energy, light, which is equal to about 186,000 miles a second.

SOUND.

BEFORE explaining the relation that sound has to telephony, it will be necessary to make quite plain what sound is, and how it affects the substance of the body through which it moves. If I strike my pencil upon the table, I hear a snap that appears to the ear to be simultaneous with the stroke: if, however, I see a man upon a somewhat distant hill strike a tree with an axe, the sound does not reach me until some appreciable time has passed; and it is noted, that, the farther away the place where a so-called sound originates, the longer time does it take to reach any listener. Hence sound has in air a certain velocity which has been very accurately measured, and found to be 1,093 feet per second when the temperature of the air is at the freezing point of water. As the temperature increases, the velocity of sound will increase a little more than one foot for every Fahrenheit degree; so that at

60° the velocity is 1,125 feet per second. This is the velocity in air. In water the velocity is about four times greater, in steel sixteen times, in pine-wood about ten times.

CONSTITUTION OF A SINGLE SOUND-WAVE.

If a person stands at the distance of fifteen or twenty rods from a cannon that is fired, he will first see the flash, then the cloud of smoke that rushes from the cannon's mouth, then the ground will be felt to tremble, and lastly the sound will reach his ear at the same time that a strong puff of air will be felt. This puff of air is the sound-wave itself, travelling at the rate of eleven hundred feet or more per second. At the instant of explosion of the gunpowder, the air in front of the cannon is very much compressed; and this compression at once begins to move outwards in every direction, so as to be a kind of a spherical shell of air constantly increasing in diameter; and, whenever it reaches an ear, the sound is perceived. Whenever such a sound-wave strikes upon a solid surface, as upon a cliff or a building, it is turned back, and the reflected wave may be heard; in which case we call it an echo. When a cannon is fired, we generally hear the sound repeated, so that it apparently lasts for a second or more; but when, as in the first case, we hear the sound of a pencil struck upon the table, but a single short report is noticed, and this, as may be supposed, consists of a single wave of condensed air.

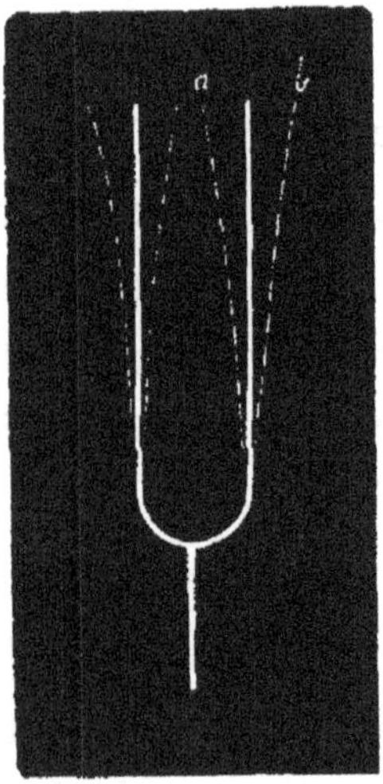

FIG. 6.

Imagine a tuning-fork that is made to vibrate. Each of the prongs beats the air in opposite directions at the same time. Look at the physical condition of the air in front of one of these prongs. As the latter strikes outwards, the air in front of it will be driven outwards, condensed; and, on account of the elasticity of the air, the condensation will at once start to travel outwards in every direction, — a wave of denser air; but directly the prong recedes, beating the air back in the contrary direction, which will obviously rarefy the air on the first side. But the disturbance we call rarefaction moves in air with the same velocity as a condensation. We must therefore remember, that just behind the wave of condensation is the wave of rarefaction, both travelling with the same velocity, and therefore always maintaining the same relative position to

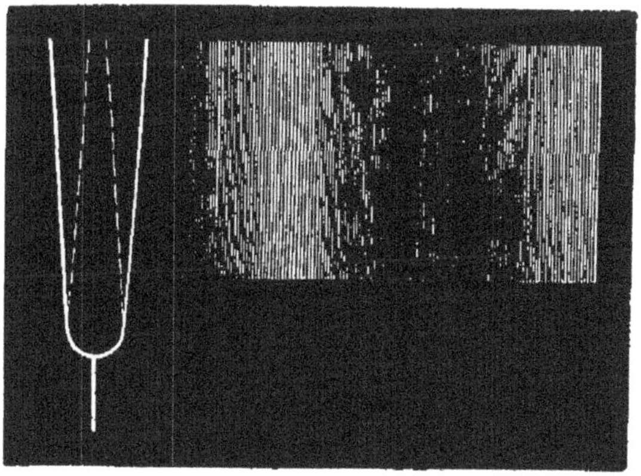

FIG. 7.

each other. Now, the fork vibrates a great many times in a second, and will consequently generate as many of these waves, all of them constituted alike, and having the same length; by length meaning the sum of the thicknesses of the condensation and the rarefaction. Suppose a fork to make one hundred vibrations per second: at the end of the second, the wave generated by the vibration at the beginning of the second would have travelled, say, eleven hundred feet; and evenly distributed between the fork and the outer limit, would be ranged the intermediate waves occupying the whole distance: that is to say, in eleven hundred feet there would be one hundred sound-waves, each of them evidently being eleven feet long. If the fork made eleven hundred vibrations per second, each of these waves would be one foot long; for sound-waves of all lengths travel in air with the same rapidity. Some late experiments seem to show that the actual amplitude of motion of the air, when moved by such a high sound as that from a small whistle, is less than the millionth of an inch.

PITCH.

The pitch of a sound depends wholly upon the number of vibrations per second that produce it; and if one of two sounds consists of twice as many vibrations per second as the other one, they differ in pitch by the interval called in music an octave, this latter term merely signifying the number of intervals into which the larger interval is divided for the ordinary musical scale. The difference between a high and a low sound is simply in the number of vibrations of the air reaching the ear in a given time. The smaller intervals into which the octave is divided stand in mathematical relations to each other when they are properly produced, and are represented by the following fractions :—

C	D	E	F	G	A	B	C
1	$\frac{9}{8}$	$\frac{5}{4}$	$\frac{4}{3}$	$\frac{3}{2}$	$\frac{5}{3}$	$\frac{15}{8}$	2

These numbers are to be interpreted thus: Suppose that we have a tuning-fork giving 256 vibrations per second: the sound will be that of the standard or concert pitch for the C on the added line as shown on the staff. Now, D when properly tuned will make 9 vibrations while C makes but 8; but, as C in this case makes 256, D must make $256 \times \frac{9}{8} = 288$. In like manner G is produced by $256 \times \frac{3}{2} = 384$, and C above by $256 \times 2 = 512$, and so on for any of the others.

If other sounds are used in the octave above or below this one, the number of vibrations of any given note may be found by either doubling or halving the number for the corresponding note in the given octave. Thus G below will consist of $\frac{384}{2} = 192$, and G above of $384 \times 2 = 768$.

During the past century there has been a quite steady rise in the standard pitch, and this has been brought about in a very curious and unsuspected way. The tuning-fork has been the instrument to preserve the pitch, as it is the best available instrument for such a purpose, it being convenient to use, and does not vary as most other musical instruments do. But a tuning-fork is brought to its pitch with a file, which warms it somewhat, so that at the moment when it is in tune with the standard that is being duplicated it is above its normal temperature; and when it cools its tone rises. When another is made of like pitch with this one, the same thing is repeated; and so it has continued until the standard pitch has risen nearly a tone higher than it was in Händel's time.

The common A and C tuning-forks to be had

in music stores, often vary a great deal from the accepted concert pitch. Such as the writer has measured have been generally too high; sometimes being ten or more vibrations per second beyond the proper number. The tuning-forks made by M. Kōenig of Paris are accurate within the tenth of one vibration, the C making 256 vibrations in one second.

LIMITS OF AUDIBILITY.

Numerous experiments have been made to determine the limits of audible sounds; and here it is found that there is a very great difference in individuals in their ability to perceive sounds. Helmholtz states that about 23 vibrations per second is the fewest in number that can be heard as continuous sound; if they are fewer in number than that, the vibrations are heard as separate distinct noises, as when one knocks upon a door four or five times a second. If one could knock evenly 23 times per second, he would be making a continuous musical sound of a very low pitch. But this limit of 23 is not the limit for all: some can hear a continuous sound with as few as 16 or 18 vibrations per second, while others are as far above the medium as this is below it. The limits of sound in musical instruments are about all included in the range of a 7-octave pianoforte from F to F, say from 42 to 5,460 vibrations per second. But this high number is not anywhere near the upper limit of audible sounds for man.

Very many of the familiar sounds of insects, such as crickets and mosquitoes, have a much higher pitch. Helmholtz puts this upper limit at 38,000 vibrations per second, and Despraetz at 36,850. The discrepancy of results is due solely to the marked difference in individuals as to acoustic perception.

For the production of high musical tones, Kōenig of Paris makes a set of steel rods. A steel rod of a certain length, diameter, and temper, will give a musical sound which may be determined. The proper length for other rods for giving higher tones may be determined by the rule that the number of vibrations is inversely proportional to the square of the length of the rod.

The dimensions of these rods when made 2 c. m. in diameter are as follows: —

Length.	Vibrations.
66.2 m. m.	20,000
59.1 " "	25,000
53.8 " "	30,000
50.1	35,000
47.5	40,000

These rods need to be suspended upon loops of silk, and they are struck with a piece of steel so short as to be wholly beyond the ability of any ear to hear its ring. Nothing but a short thud is to be heard from it when it strikes, while from the others comes a distinct ringing sound. In experimenting with such a set of steel rods I have not found any one yet who could hear as many as 25,000 per second, my own limit being about 21,000. But it has been experimentally found that children and youth have a perceptive power for high sounds considerably above adults. Dr. Clarence Blake of Boston reports a case in his aural practice, of a woman whose hearing had been gradually diminishing for some years until she could not hear at all with one ear, and the ticking of a watch could only be heard with the other when the watch was held against the ear.

After treatment it was discovered that the sensibility to high sounds was very great, and that she could hear the steel rod having a tone of 40,000 vibrations.

Last year Mr. F. Galton, F.R.S., exhibited before the Science Conference an instrument in the shape of a very small whistle, which he had devised for producing a very high sound. The whistle had a diameter less than the one twenty-fifth of an inch. The length could be varied by moving a plug at the end of the whistle. It was easy to make a sound upon such an instrument that was altogether out of hearing-range of any person. Mr. Galton tried some very interesting experiments upon animals, by using these whistles. He went through the Zoölogical Gardens, and produced such high sounds near the ears of all the animals. Some of them would prick up their ears, showing that they heard the sound; while others apparently could not hear it. He declares that among all the animals the cat was found to hear the sharpest sound. Small dogs can also hear very shrill notes, while larger ones can not. Cattle were found to hear higher sounds than horses. The squeak of bats and of mice cannot be heard by many persons who can hear ordinary sounds as well as any; sharpness of hearing having nothing to do with the limits of hearing.

EFFECTS OF SOUND UPON OTHER BODIES.

If a vibrating tuning-fork be held close to a delicately suspended body, the latter will approach the fork, as if impelled by some attractive force. The experiment can be made by fastening a bit of paper about an inch square to a straw five or six inches long, and then suspending the straw to a thread, so that it is balanced horizontally. Bring the vibrating tuning-fork within a quarter of an inch of the paper. In this case the motion of approach is due to the fact that the pressure of the air is less close to a vibrating body than at a distance from it; there is therefore a slightly greater pressure on the side of the paper away from the fork than on the side next to it.

If a vibrating tuning-fork be held near to the ear, and turned around, there may be found four places in one rotation where the sound will be heard but very faintly, while in every other position it can be heard plainly enough. The extinction of the sound is due to what is called interference. Each of the prongs of the fork is giving out a sound-wave at the same time, but in opposite directions, each wave advancing outwards in every direction. Where the rarefied part of one wave exactly balances the condensed part of the other, there of course the sound will be extinguished; and these lines of interference are found to be hyperbolas, or, if considered with reference to both entire waves, two hyperbolic surfaces.

SYMPATHETIC VIBRATIONS.

When it is once understood that a musical sound is caused by the vibrations more or less frequent which only make the difference we call pitch, it might at once be inferred, that if we have a body that is capable of vibrating say a hundred times a second, and it receives a hundred pulses or pushes a second, it would in this way be made to vibrate. Suppose, then, that we take two tuning-forks, each capable of vibrating 256 times a second: if one be struck while the other is left free, the former one will be giving to the air 256 impulses per second, which will reach the other fork, each pulse tending to move it a little, the cumulative result being to make it move perceptibly, that is, to give out a sound. The principle is just the same as that employed in the common swing. One push makes the swing to move a little, upon its return another is given, in like manner a third, and so on until a person may be swung many feet high. If a glass tumbler be struck, it gives out a musical sound of a certain pitch, which will set a piano-string sounding that is tuned to the same pitch, provided that the damper be raised. It is said that some persons' voices have broken tumblers by singing powerfully near them the same note which the tumblers could give out, the vibrations of the tumblers being so great as to overcome cohesion of the molecules.

There are very many interesting effects due to sympathetic vibrations.

Large trees are sometimes uprooted by wind

that comes in gusts timed to the rate of vibration of the tree. When troops of soldiers are to cross a bridge, the music ceases, and the ranks are broken, lest the accumulated strain of timed vibrations should break the structure; indeed, such accidents have several times occurred. There is not so much danger to a bridge when it is heavily loaded with men or with cattle, as when a few men go marching over it. "When the iron bridge at Colebrooke Dale was building, a fiddler came along, and said to the workmen that he could fiddle their bridge down. The builders thought this boast a fiddle-de-dee, and invited the musician to fiddle away to his heart's content. One note after another was struck upon the strings, until one was found with which the bridge was in sympathy. When the bridge began to shake violently, the workmen were alarmed at the unexpected result, and ordered the fiddler to stop."

Some halls and churches are wretchedly adapted to hear either speaking or singing in. If wires be stretched across such halls, between the speaker's stand and the opposite end, they will absorb the passing sound-waves, and will be made to sympathetically vibrate, thus preventing in a good degree the interfering echoes. The wire should be rather fine piano-wire, and it should be stretched so tightly as to give out a low musical sound when plucked with the fingers. In a large hall there should be twenty or more such wires.

RESONANCE.

When a tuning-fork is struck, and held out in the air, the vibrations can be felt for a time by the fingers; but the sound is hardly audible unless the fork be placed close to the ear. Let the stem of the fork rest upon the table, a chair, or any solid body of considerable size, and the sound is so much increased in loudness as to be heard in every part of a large room. The reason appears to be, that in the first case the vibrations are so slight that the air is not much affected. Most of the force of the vibration is absorbed by the hand that holds it; but when the stem rests upon a hard body of considerable extent, the vibrations are given up to it, and every part of its surface is giving off the vibrations to the air. In other words, it is a much larger body that is now vibrating, and consequently the air is receiving the amplified sound-waves.

If the stem of the fork had been made to rest upon a bit of rubber, the sound would not only not have been re-enforced in such a way, but the fork would very soon have been brought to rest; for India rubber *absorbs* sound vibrations, and converts them into heat vibrations, as is proved by placing such a combination upon the face of a thermo-pile.

If one will but put his hand upon a table or a chair-back in any room where a piano or an organ is being played, or where voices are singing, especially in church, he cannot fail to feel the sound; and if he notices carefully he will perceive that some sounds make such table or seat to shake much more vigorously than others, — a genuine case of sympathetic vibrations.

It is for this reason that special materials and shapes are given to parts of musical instruments, so that they may respond to the various vibrations of the strings or reeds. For instance, the piano has an extensive thin board of spruce underneath all the strings, which is called the sounding-board. This board takes up the vibrations of the strings ; but, unlike the rubber, gives them all out to the air, greatly re-enforcing their strength, and changing somewhat their quality. But the air itself may act in like manner. In almost any room or hall not more than fifteen or twenty feet long, a person can find some tone of the voice that will seem to meet some response from the room. Some short tunnels will from certain positions yield very powerful, responsive, resonant tones. There is certainly one such in Central Park, New York. It is forty or fifty feet long. To a person standing in the middle of this, and speaking or making any kind of a noise on a certain pitch, the resonance is almost deafening. It is easy to understand. When a column of air enclosed in a tube is made to vibrate by any sound whose wave-length is twice the length of the tube, we have such column of air now filled with the condensed part of the wave, and now with the rarefied part ; and as these motions cannot be conducted laterally, but must move in the direction of the length of the tube, the air has a very great amplitude of motion, and the sound is very loud. If one end of the tube be closed, then the length must be but one-fourth of the wave-length of the sound. Take a tuning-fork of any convenient pitch, say a C of 512 vibrations per second: hold it while vibrating over a vertical test-tube about eight inches long. No response will be heard; but, if a little water be carefully poured into the tube to the depth of about two inches, the tube will respond loudly, so that it might be heard over a large hall. In this case the length of the air-column that was responding, being one-fourth the wave-length, would give twenty-four inches as the wave-length of that fork.

It is easy in this way to measure approximately the number of vibrations made by a fork.

Letting $l =$ depth of tube,
$d =$ diameter of tube,
$v =$ velocity of sound reduced for temperature,
$N =$ number of vibrations,

Then $N = \dfrac{v}{4\,(l + d)}$.

When a vibrating tuning-fork is placed opposite the embouchure of an organ-pipe of the same pitch, the pipe will resound to it, giving quite a volume of sound. In 1872 it occurred to me, that the action of an organ-pipe might be quite like that of a vibrating reed in front of the embouchure. As the air is driven past it from the bellows, the form of the escaping air will evidently be like a thin, elastic strip; and, having considerable velocity, it will carry off by friction a little of the air in the tube: this will of course rarefy the air in the tube somewhat, and a wave of condensation will travel down the tube. At the bottom, being suddenly stopped, its re-action will be partly outwards, and so will drive the strip of air away from the tube. After this will follow, for a like reason, the other phase of the wave, the rarefaction, which will swing the strip of air towards the tube. This theory I verified by filling the bellows with smoke, and watching the motion of the escaping air and smoke with a stroboscope. This view is now advocated by an organ-builder in England, Herman Smith; but whether he discovered it before or after me, I do not know.

When a membrane vibrates, its motion is generally perceptible to the eye; and it may have a very great amplitude of motion, as in the case of the drum; and various instruments have been devised for the study of vibrations, using membranes like rubber, gold-beater's skin, or even tissue paper, to receive the vibrations. One of the musical instruments of a former generation of boys was the comb. A strip of paper was placed in front of it, and placed at the mouth, and sung through, the paper responding to the pitch with a loose nasal sound. Köenig fixed a membrane across a small capsule, one side of which was connected by a tube to any source of sound, and the other side to a gas-pipe and a small burner. A sound made in the tube would shake the flame, and a mirror moving in front of the flame would show a zigzag outline corresponding to the sound vibrations.

In like manner if a thin rubber be stretched over the end of a tube one or two inches in diameter and four or five inches long, and a bit of looking-glass one-fourth of an inch square be made fast to the middle of the membrane, the motions of the latter can be seen by letting a beam of sunlight fall upon the mirror so as to be reflected upon a white wall or screen a few foot away. (Fig. 8.)

When a sound is made in this tube, the spot of light will at once assume some peculiar form,—either a straight line with some knots of light in

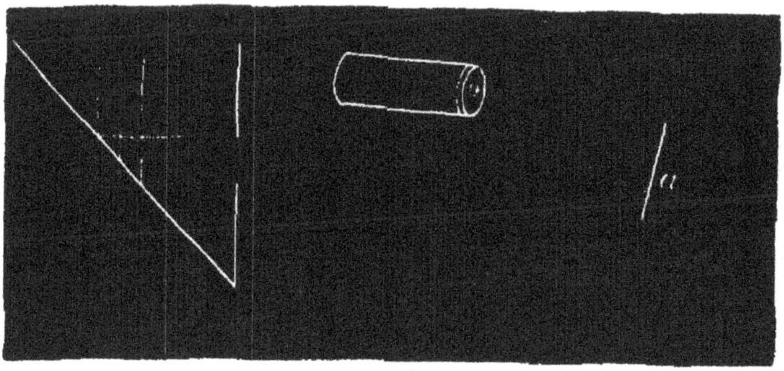

FIG. 8.

it, or some curve simple or compound, and such as are known as Lissajous curves. If, while some of these forms are upon the screen, the instrument be moved sideways, the forms will change to undulating lines with or without loops, varying with the pitch and intensity, but being alike for the same pitch and intensity. (Fig. 9.)

This instrument I called the opeidoscope.

The vibration of a membrane and that of a solid differ chiefly in the amplitude of such vibration. The scratch of a pin at one end of a long log can be heard by an ear applied to the other end of the log; but every molecule in the log must move slightly; and there are all degrees of movement between that visible to the eye, which we call mass motion, and that called molecular simply because we cannot measure the amplitude of the motion. We may, then, roughly divide all bodies into two classes, as to their relations to sound, — such as re-enforce it, and such as distribute it: the first depending upon the form of the body, as related to a particular sound; the second independent of form, and responding to all orders of vibrations. Air, wood, and metals belong in this latter class. The common toy-string telegraph, or *lovers' telegraph*, is an example

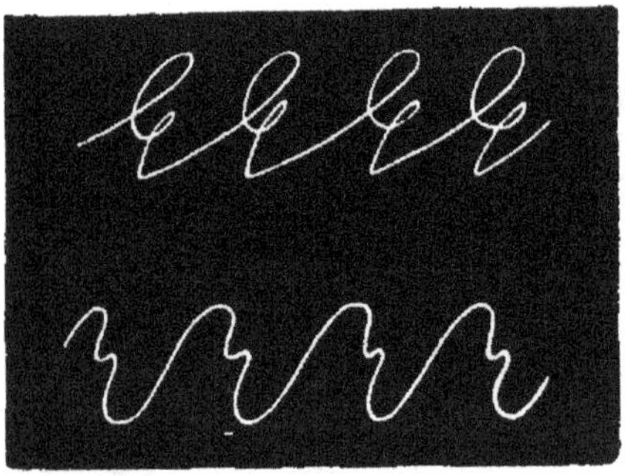

FIG. 9.

of this class. Two tin boxes are connected by a string passing through the middle of the bottom of each. When the string is stretched, and a person speaks in one box, what is said can be heard by an ear applied at the other. If the speaking-tubes be made about four inches in diameter, and about four inches deep, they are capable of doing much more service than is generally supposed to be possible. I know of two lines, one of five hundred feet and the other of a thousand feet in length, over which one can talk, and be heard with distinctness. In the line of a thousand feet, the end of the tube is made of sheepskin tightly stretched, and the line is made of No. 8 cotton thread. The greater the tension, the better is the sound transmitted. The thread is supported at intervals by running through a loop on the ends of cords not less than three feet long, attached to supports. The thread pierces the membrane, and is attached to a small button which is in contact with the membrane. Wind and rain affect this line disadvantageously. The other line of five hundred feet, between a passenger and a freight depôt, has the tube end covered with stretched calfskin. Instead of thread, a copper-relay wire is employed (any small uninsulated wire will do as well). This permits a good tension, and is unaffected by the weather. One may stand in front of it about three feet, and converse with ease, and in an ordinary tone. The wire is supported in loops of string, as in the other.

Musicians have in all times employed various instruments for the production of musical effects. Whistles made of bone were used by pre-historic men, some of them having finger-holes so that different tones could be produced. A stag-horn that was blown like a flageolet, and having three finger-holes, has also been found; while on the old monuments of Egypt are pictured harps, pipes with seven finger-holes, a kind of flute, drums, tambourines, cymbals, and trumpets. In later times these primeval forms have been modified into the various instruments in use in the modern orchestra. It seems as if no musician had ever been interested in the question as to why one instrument should give out a sound so different from another one, even though it was sounding upon the same pitch. No one can ever mistake the sound of a violin, or a horn, or a piano, for any other instrument; and no two persons have voices alike. This difference in tone, which enables us to identify an instrument by its sound or a friend by his voice, is called quality of tone, or *timbre.*

About twenty years ago, that great German physicist Helmholtz undertook the investigation of this subject, and succeeded in unravelling the whole mystery of the qualities of sound.

He discovered first, that a musical sound is very rarely a simple tone, but is made up of several tones, sometimes as many as ten or fifteen, having different degrees of intensity and pitch. The lowest sound, which is also the strongest, is called the *fundamental;* and it is this tone we mean when we speak of the pitch of a sound, as the pitch of middle C upon a piano, or the pitch of the *A* string on a violin. The higher sounds that accompany the fundamental are called sometimes harmonics, sometimes upper partial tones, but generally *overtones.* The character or quality of a sound depends altogether upon the number and intensity of these overtones associated with the fundamental. If a sound can be made upon a pipe and a violin, that consists wholly of the fundamental with no overtones, the two instruments sound absolutely alike. It is exceedingly difficult to do this; and such sound when produced is smooth, but without character, and unpleasing.

Second, Helmholtz discovered that the overtones always stand in the simplest mathematical relation to the fundamental tone, — in fact, are simple multiples of that tone, being two, three, four, and so on, times the number of vibrations of it.

This will be readily understood by considering the position of such related sounds when they are written upon the staff.

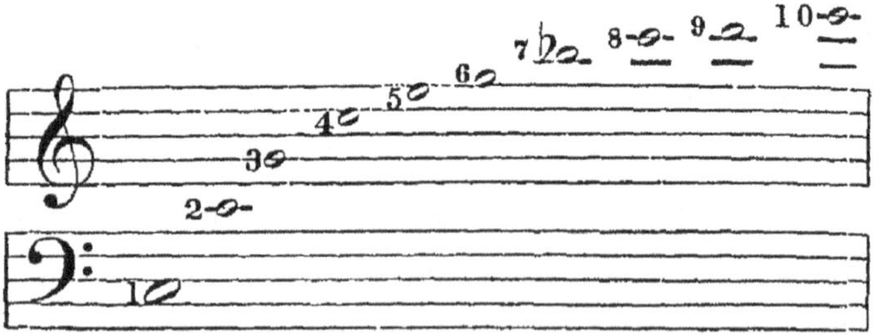

If we start with C in the bass as indicated in the staff, calling that the fundamental, then the notes that will represent the above ratios are those indicated by smaller notes, which are the overtones up to the ninth. The first overtone, being produced by twice the number of vibrations, must be the octave; the second, the fifth of the second octave; the third will be two octaves from the first, and so on: the number of vibrations of each of these notes being the number of the fundamental multiplied by its order in the series.

Taking C with 128 vibrations, we have for this series:—

128 × 1 = 128 = C fundamental.
128 × 2 = 256 = C′.
128 × 3 = 384 = G′.
128 × 4 = 512 = C″.
128 × 5 = 640 = E″.
128 × 6 = 768 = G″.
128 × 7 = 896 = B″♭.
128 × 8 = 1,024 = C‴.
128 × 9 = 1,152 = D‴.
128 × 10 = 1,280 = E‴.

This series is continued up to the limits of hearing. Now, it appears that all instruments do not give the complete series: indeed, it is not possible to obtain them all upon some instruments. Each of them, however, when present helps in the general effect which we call quality. Sometimes the overtones are more prominent than the fundamental, as when a piano-wire is struck with a nail. It has always been noticed that it does not give out the sound that is wanted when it is struck in this way. Hence it is the art of an instrument-maker to so construct the instrument as to develop and re-enforce such tones as are pleasing, and to suppress the interfering and disagreeable overtones. Piano-makers learned by trial where was the proper place to strike the stretched wire in order to develop the most musical sound upon it; but no reason could be given until it was observed that striking it at a point about one-seventh or one-ninth its length from either end prevented the development of the objectionable overtones, the seventh and the ninth. Hence they can scarcely be heard in a properly constructed instrument. These overtones are very discordant with the lower sounds.

Organ-pipes have their specific qualities given

to them by making them wide-mouthed, narrow-mouthed, conical, and so on; shapes which experience has determined give pleasing sounds with different qualities.

The violin is an instrument that seems to puzzle makers more than almost any other. Some of the old violins made two hundred years ago by the Amati family at Cremona are worth many times their weight in gold. Recent makers have tried in vain to equal them; but, when their ingenuity and skill have failed, they declare that *age* has much to do with such instruments, that age mellows the sounding quality of the violin. But the Cremona violins were just as extraordinary instruments when they left the hands of the makers as they are now; and the fame of the Amati family as violin-makers was over all Europe while they were living.

A good violin when well played gives an exquisite musical effect, and on account of its range and quality of tones it is the leading orchestral instrument, always pleasing and satisfying; but in unskilled hands even the best *Cremona* will give forth sounds that make one grieve that it was ever invented. Overtones of all sorts and with all degrees of prominence may be easily developed upon it: therefore the skilful player draws the bow at such a place upon the strings as to develop the overtones he wants, and suppress the ones not wanted. The usual rule is to draw the bow about an inch below the bridge; but the place for the bow depends upon where the fingers are that stop the strings, and also the pressure upon it. It requires an almost incredible amount of practice to be able to play a violin very well.

In the accompanying table will be found the component parts of tones upon a few instruments in common use.

TONE COMPOSITION.

The components of the tones are indicated by lines in the column underneath the figures representing the series. Thus the narrow-stopped organ-pipe gives a sound composed of a fundamental, and overtones three, five, seven, and nine times the number of vibrations of it.

It must not be inferred that all of the overtones are of equal strength: they are very far from

TONE COMPOSITION.

	INSTRUMENTS.	1	2	3	4	5	6	7	8	9	10
ORGAN PIPES.	Wide stopped..	/									
ORGAN PIPES.	Narrow "	/		/		/		/		/	
ORGAN PIPES.	Narrow cylinder	/	/	/	/	/	/				
ORGAN PIPES.	Principal (Wood)	/	/	/							
ORGAN PIPES.	Conically narrow at top.	/				/	/	/			
	Flute.............	/	/	/	/						
	Violin	/	/	/	/	/	/	/	/	/	/
	Piano.............	/	/	/	/	/	/	/	/		
	Bell	/	/	/	/	/	/	/			
	Clarionet..........	/		/		/		/		/	
	Bassoon...........	/	/	/	/	/	/	/			
	Oboe	/	/	/	/	/	/	/			

that; but these differ in different instruments, and it is this that constitutes the difference between a good instrument and a poor one of the same name.

In a few of the spaces very light lines are made for the purpose of indicating that such overtones are quite weak. For instance: the piano has the sixth, seventh, and eighth thus marked; these tones being suppressed by the mechanism, as described on a former page.

Only a few of the many forms of organ-pipes are given; but these are sufficient to show what a physical difference there is between the musical tones in such pipes.

As for the human voice, it is very rich in overtones; but no two voices are alike, therefore it would be impossible to tabulate the components of it in the manner they are tabulated for musical instruments.

In Helmholtz's experiments in the analysis of sounds, use was made of the principle of resonance of a body of air enclosed in a vessel. In the experiment with the tuning-fork to determine the wave-length, p. 78, it is remarked that no response came until the volume of the air in the tube was reduced to a certain length, which depended upon the vibration number of the fork. If instead of a test-tube a bottle had been taken, the result would have been the same. Every kind of a vessel can respond to some tone of a definite wave-length, and a sphere has been found to give the best results. These are made with a hole on one side for the sound-wave to enter, and a projection on the opposite side, through which a hole about the one-eighth of an inch is made, this to be placed in the ear. Any sound that is made in front of the large orifice will not meet any response, unless it be that particular one which the globe can naturally re-enforce, when it will be plainly heard. Suppose, then, one has a series of twenty or more of these, graduated to the proper size for re-enforcing sounds in the ratio of one, two, three, four, and so on. Take any instrument, say a flute: have one to blow it upon the proper pitch to respond to the largest sphere, then take each of the spheres in their order, applying them to the ear while the flute is being sounded. When the overtones are present they will be heard plainly and distinct from the fundamental sound. In like manner any or all other sounds may be studied.

But Helmholtz did not stop after analyzing sounds of so many kinds: he invented a method of synthesis, by which the sounds of any kind of an instrument could be imitated. A tuning-fork, when made to vibrate by an electric current, gives out a tone without harmonics or overtones. So if a series of forks with vibration periods equal to the numbers of the series of overtones given on p. 86 be so arranged that any of them may be made to vibrate at will, it is evident that the resulting compound tone would be comparable with that from an instrument having such overtones. Thus, if with a tuning-fork giving a fundamental C, other forks giving two, three, and four times the number of the fundamental were associated, each one giving a simple tone, we should have for a resultant the tone of a flute, as shown on p. 91. If one, three, five, seven, and nine, were all sounded, the resulting tone would be that of the clarionet, and so on. This he actually accomplished, and now makers of physical apparatus advertise just such instruments.

Helmholtz also contrived a set of tuning-forks, which, when bowed, will give out the vowel sounds like the voice.

It was remarked upon p. 89 that it has generally been considered that age has a mellowing effect upon the sound of a violin. Once in possession of the facts concerning sound that have been alluded to on the preceding pages, it is easy to see how such an opinion should arise, and also the fallacy of it. It is proved conclusively that the ability to hear high sounds decreases as one grows older. As the violin gives a very great number of overtones, even up to the limits of audibility, it is plain that if such an instrument should not change in its quality of tone in the least degree, yet to a man who played upon it for a number of years it would seem to change by subtracting some of the higher overtones from the sound; that is, it would seem to become mellower. There is no evidence that such a physical change takes place in the instrument. It is not here affirmed that no change does take place. It may be probable; but all the evidence we have is the opinions of individuals whose hearing we know does change; and this change is competent to modify the judgment as to the quality of the sound in the same direction. Before it can be affirmed that such a physical change does take place in the violin as to make a perceptible difference in the quality of its tone, it will be needful to determine accurately the number and intensity of the overtones at intervals during many years, and then to compare them. This has not yet been done.

FORM OF A COMPOUND SOUND-WAVE IN AIR.

Upon p. 63 is given a picture of the form of a simple sound-wave in air, which, as described, consists of two parts, a condensation and a rarefaction. All simple sound-waves have such a form; but when two or more sound-waves that stand in some simple ratio to each other, as do the sounds of musical instruments, are formed in air, the resulting wave is more or less complex in structure; and where there are many components, as there are where a number of different kinds of instruments are all sounding at once, it is well-nigh impossible to figure even approximately the form of such wave-combinations. It is generally given in treatises upon sound with ordinates representing the factors with their relative intensities. When the extremities of the ordinates are connected, there is drawn a curved line with regularly recurring loops. This cannot give a correct idea of the form of the wave, because the motion of a particle of air is not up and down like a floating body upon waving water, but it is forward and back, in the direction of the motion of the wave.

In Fig. 10 three simple sound-waves are thus represented at 1, 2, and 3, these having the wave-length 1, 2, and 3. In 4, the three are combined into one compound wave, and better show the form of a transverse section of such a sound-wave in the air. The organ-pipe called the principal gives out such a compound wave as is seen by referring to the table on p. 91. The second overtone, however, is quite weak in that pipe, which would so modify the form as to lessen somewhat the density at *b*, and increase it at *a*.

In like manner the space in the length of the fundamental sound, whatever it may be, is divided up into a number of minor condensa-

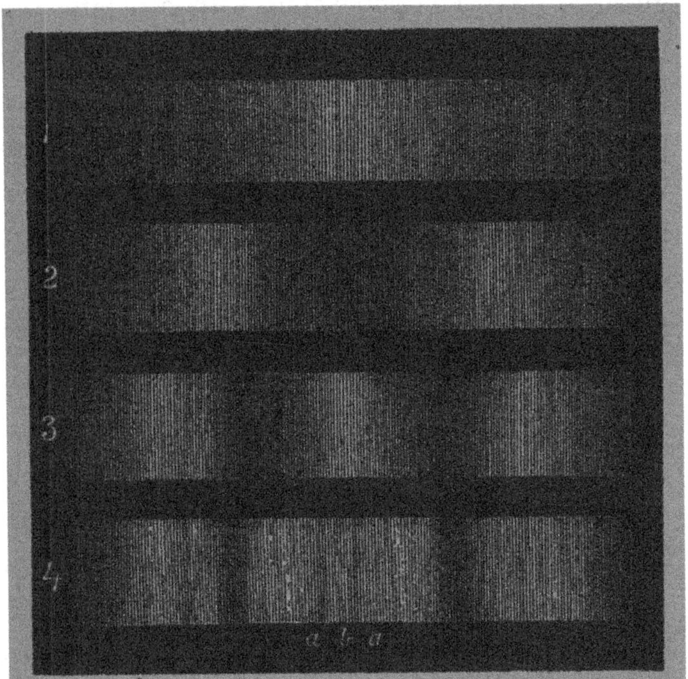

FIG. 10.

tions and rarefactions, which may strengthen each other, or so interfere as to change the position of both; as is seen in the figure at *b*, where the condensation due to wave 2 interferes with the rarefaction of 3.

CORRELATION.

Having treated at some length of the three factors involved in telephony, — namely, electricity, magnetism, and sound, — it remains to follow up the various steps that have led to the actual transmission of musical sounds and speech over an ordinary electric circuit.

It is stated upon p. 31, that, when a current of electricity is passed through a coil of wire that surrounds a rod of soft iron, the latter is made a temporary magnet: it loses its magnetic property the instant that the current ceases. If the rod be of considerable size, say a foot or more in length, and half an inch or more in diameter, and the current be strong enough to make a powerful magnet of it, whenever the current from the battery is broken, the bar may be heard to give out a single *click*. This will happen as often as the current is broken. This is occasioned by a molecular movement which results in a *change* of *length* of the bar. When it is made a magnet, it elongates about $\frac{1}{25000}$ of its length; and, when it loses its magnetism, it *suddenly* regains its original length; and this change is accompanied with the sound. This sound was first noticed by Prof. C. G. Page of Salem, Mass., in 1837. If some means be devised for breaking such a circuit more than fifteen or sixteen times a second, we shall have a continuous sound with a pitch depending upon the number of clicks per second. Such a device was first invented by the same man, and was accomplished by fixing the armature of an electro-magnet to a spring which was in the circuit when the spring was pressing against a metallic knob, at which time the current made the circuit in the coil of the electro-magnet. The magnet attracting the armature away from the button broke the circuit, which of course destroyed the magnetism of the magnet, and allowed the spring to fly back against the button, to complete the circuit and reproduce the same series of changes. The rapidity with which the current may be broken in this way is only limited by the strength of both spring and current. The greater the tension of the spring with a given current, the greater number of vibrations will it make.

Suppose such an intermittent current to pass through the coil surrounding the soft iron rod, 256 times per second; then the rod would evi-

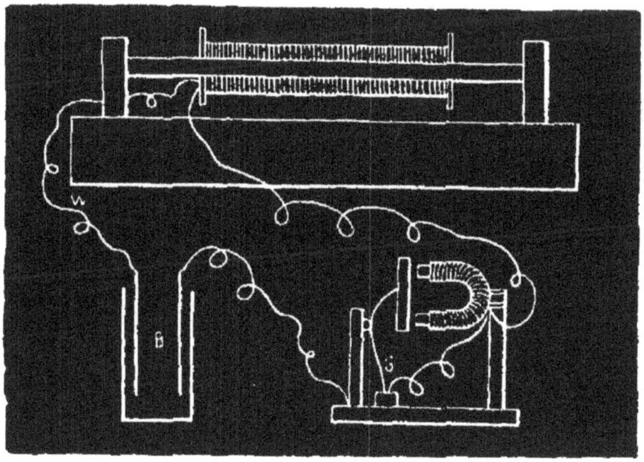

FIG. 11.

dently give 256 clicks per second, which would have the pitch of C. When these clicks are produced in the rod held in the hand, the sound is hardly perceptible, being like that of a sounding tuning-fork when held thus. In order to strengthen it, it is necessary to place it on some resonant surface. It is customary to mount it upon an oblong box with one or two holes in its upper surface, inasmuch as such a form is found to give a louder response than any other, and is the shape usually given to Æolian harps. The accompanying cut shows the combination of battery B, the circuit-breaker, and the rod mounted upon the box. The wire W may evidently be of any length, the magnetized rod and box responding to the number of vibrations of the spring S, how long soever the circuit may be.

HELMHOLTZ' ELECTRIC INTERRUPTOR.

In some of Helmholtz' experiments, it was essential to maintain the vibrations of a tuning-fork for a considerable time. He effected this by placing a short electro-magnet between the prongs of the fork, and affixing a platinum point at the end of one prong in such a manner, that, as the prong descended in its vibration, the platinum point dipped into a small cup of mercury that completed the circuit. When the prong receded, it was of course withdrawn from the mercury, and the current was broken. As it is not possible for a tuning-fork to vibrate in more than one period, such an arrangement would evidently make and break the current as many times per second as the fork vibrated. When, therefore, such an interruptor is inserted in the circuit with the click-rod on its resonant box, the latter must give out just such a sound as the fork is giving. With such a device, it is possible to reproduce at almost any distance in a telegraphic circuit, a sound of a given pitch. It is therefore a true telephone.

REISS' TELEPHONE.

The ease with which membranes are thrown into vibrations corresponding in period to that of the sounding body has already been alluded to on p. 80; and several attempts have been made, at different times, to make membranes available in telephony. The first of these attempts was made by Philip Reiss of Friedrichsdorf, Germany, in 1861.

His apparatus consisted of a hollow box, with two apertures: one in front, in which was inserted a short tube for producing the sound in, and indicated by the arrow in the cut, Fig. 12;

the other on the top. This was covered with the membrane *m*, — a piece of bladder stretched tight over it. Upon the middle of the membrane, a thin piece of platinum was glued; and this piece of platinum was connected by a wire

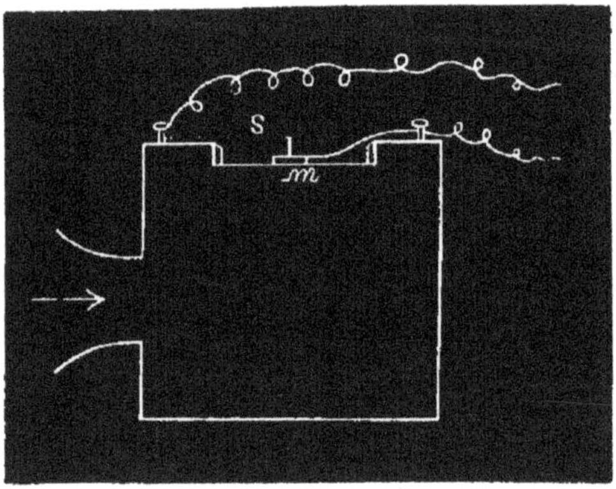

FIG. 12.

to a screw-cup from which another wire went to a battery.

A platinum finger, S, rested upon the strip of platinum, but was made fast at one end to the screw cup that connected with the other wire from the battery. Now, when a sound is made in the box, the membrane is made to vibrate powerfully: this makes the platinum strip to strike as often upon the platinum finger, and as often to bound away from it, thus making and breaking the current the same number of times per second. If, then, a person sings into this box while it is in circuit with the afore-mentioned click-rod and box, the latter will evidently change its pitch as often as it is changed by the voice. In this apparatus we have a telephone with which a melody may be reproduced at a distance with distinctness. But the sounds are not loud, and they have a tin-trumpet quality. If one reflects upon the possibilities of such a mechanism, and upon the conditions necessary to produce a sound of any given quality, as that of the voice or of a musical instrument as described in preceding pages, he will understand that it can reproduce only pitch. It might here be inferred that something more than a single pitch is transmitted if the sound is like that of a tin trumpet as stated: but the reason of this is that, whenever a current is passing between two surfaces that can move only slightly on each other, there is always an irregularity in the conduction, so as to produce a kind of scratching sound; and it is this, combined with the other, the true pitch, that gives the character to the sound of this instrument.

Dr. Wright found that a sound of considerable intensity could be obtained by passing the interrupted current through the primary wire of a small induction coil, and placing a conductor made of two sheets of silvered paper placed back to back in the secondary circuit. The silvered paper becomes rapidly charged and discharged, making a sound that can be heard over a large hall, and having the same pitch as the sending instrument.

GRAY'S TELEPHONES.

In 1873 Mr. Elisha Gray of Chicago discovered that if an induction coil be made to operate by the current from any automatic circuit-breaker, and one of the wires from the secondary circuit be held in the hand while the dry finger of the same hand is rubbed upon a sonorous metallic plate, the other wire being in connection with the plate, a musical sound would be given outby the plate, appearing to come from the point of contact of the finger with the plate. He therefore contrived a musical instrument with a range of two octaves, in which the reeds were made to vibrate by electro-magnets, the current entering any one by depressing the appropriate key. This circuit is sent through the primary wire of an induction coil while one of the terminals of the secondary coil is connected with the thin sheet metal that forms one head of a shallow wooden drum about eight inches in diameter, which is so fixed as to be rotated like a pulley. The other terminal is held in the hand while one finger of the same hand rests upon the metallic surface. While the drum is turned with the other hand, the sounds given out have considerable intensity. The faster the drum is turned, the louder do the sounds become, though the pitch remains the same.

In this case, as in the case mentioned on p. 105, we have an electric current passing between two surfaces that are moving upon each other; the contact not being uniform, the current is varying as well as intermittent.

Mr. Gray has also invented a musical telephone by means of which many musical sounds may be simultaneously transmitted and reproduced. The actual mechanism used is quite complex, and requires considerable familiarity with electrical science in order to understand it; but the fundamental principle involved is not difficult to one who has comprehended the preceding descriptions.

Suppose that we have a series of four steel reeds, each one fixed at one end to one pole of a short electro-magnet, while the other end is left free to vibrate over the other pole of the magnet and not quite touching it. Each of the reeds is to be tuned to a different pitch, say the 1, 3, 5, and 8 of the scale. These electro-magnets with their attached vibrators are to be attached each to a resonant box (see p. 93), which can respond to that particular number of vibrations per second. This is the receiving instrument. The sender consists of a like set of reeds tuned to the same pitch, which can be made to vibrate at will by pressing a key which sends the current of electricity through its electro-magnet, which makes and breaks the current. Imagine one of these keys to be pressed down so as to make the circuit complete: the sending instrument then has one of its reeds, let it be the 1 of the scale, set in vibration; the intermittent current traverses the whole line, going through all four of the receiving instruments. Now, we know from the study of the action of sounding bodies, that only one of the four receivers is competent to vibrate in consonance with this tone, and this one will respond; that is, the vibrations are truly sympathetic vibrations. If, instead of making the 1 of the scale in the sending-instrument, the 3 had been made, the current would have gone through all of the receiving instruments just the same as before, but only one of them could take up that vibratory movement: three of them would remain at rest, the 3 responding loudly. In like manner, any number of vibrating reeds in the sending instrument can make a corresponding number of reeds in the receiving instrument to vibrate, provided the latter be exactly tuned with the former. Each transmitter is connected with but a part of the battery, so that several tones may be transmitted at the same time. If the performer plays a piece of music in its various parts, every part will be reproduced: thus we have a compound or multiple telephone. This instrument has been used during the past winter to give concerts in cities when the performer was in a distant place.

It has also been used as a multiple telegraph; as many as eight operators sending messages simultaneously over the same wire, — four in each direction, — without the slightest interference.

BELL'S TELEPHONE.

Prof. A. Graham Bell of Boston independently discovered the same means for producing multiple effects over the same wire; but it appears he did not practically work it out as completely as did Mr. Gray. But while the latter was chiefly employed in perfecting the method as a telegraphic system, Prof. Bell had set before himself the more difficult problem of transmitting speech. This he has actually accomplished, as we have so often been reminded during the past year.

Thoroughly conversant with the acoustic researches of Helmholtz, and keeping in mind the complex form of the air vibrations produced by the human voice, he attempted to make these vibrations produce corresponding pulsations in an electric current in the manner analogous to the electric interruptor.

Observing that membranes when properly stretched can vibrate to any kind of a sound, he sought to utilize them for this purpose. So did Reiss; but Reiss inserted the vibrating membrane into the circuit, and it was quite evident that such a plan would not answer, therefore the current must not be broken; but could an electric current be interfered with without breaking the connections?

The well-known re-actions of magnets upon electrical currents, first noted by Oersted, and fully developed by Faraday, gave the clew to the solution. A piece of iron should be made to vibrate by means of sound vibrations, so as to affect an electro-magnet and induce corresponding electrical pulsations.

FIRST FORM OF SPEAKING-TELEPHONE.

A membrane of gold-beater's skin was tightly stretched over the end of a speaking-tube or funnel; on the middle of this membrane a piece of iron, N S, Fig. 13, was glued. In front of this piece of iron an electro-magnet M is so situated that its poles are opposite to it, but not quite touching it. One of the terminal wires of the electro-magnet goes to the battery B; the other goes to the receiving instrument R, which consists of a tubular electro-magnet, the coil being enclosed in a short tube of soft iron; the wire thence goes to the plate E′, which is sunk in the earth. On the top of R, at

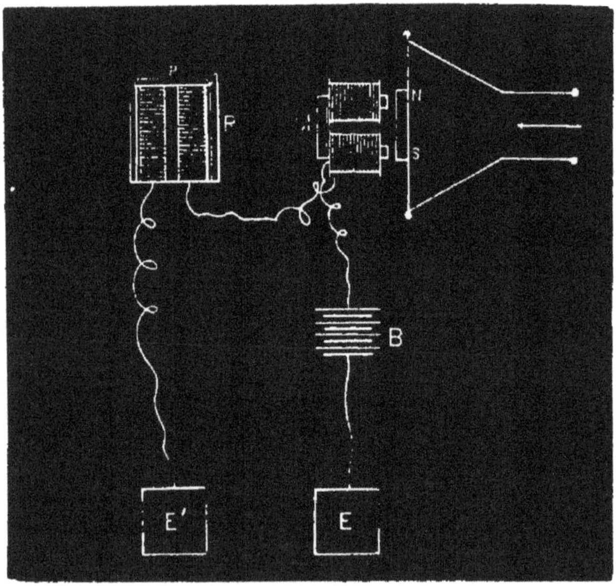

FIG. 13.

P, is a rather loose, thin disk of iron, which acts as an armature to the electro-magnet below it.

Supposing that all the parts are thus properly connected, the current of electricity from the battery makes both M and R magnetic; the electro-magnet M will inductively make the piece of iron N S, a magnet, with its poles unlike those of the inducing electro-magnet; and the two will mutually attract each other. If now this piece of iron N S be made to move toward M, a current of electricity will be induced in the coils, which will traverse the whole circuit. This induced electricity will consist of a single wave or pulse, and its force will depend upon the velocity of the approach of N S to M. A like pulse of electricity will be induced in the coils when N S is made to move away from M; but this current will move through the circuit in the opposite direction, so that whether the pulsation goes from M to R, or from R to M, depends simply upon the direction of the motion of N S.

The electricity thus generated in the wire by such vibratory movements varies in strength proportional to the movement of the armature; therefore the line wire between two places will be filled with electrical pulsation exactly like the aërial pulsations in structure. Fig. 10, p. 98, may be used to illustrate the condition of the wire through which the currents pass. The dark part may represent the strongest part of the wave, while the lighter part would show the weaker part of the wave. The chief difference would be, that electricity travels so fast, that what is there represented as one wave in air with a length of two feet would, in an electric wave, be more than fifty miles long.

These induced electric currents are but very transient (see p. 31); and their effect upon the receiver R is to either increase or decrease the power of the magnet there, as they are in one direction or the other, and consequently to vary the attractive power exercised upon the iron plate armature.

Let a simple sound be now made in the tube, consisting of 256 vibrations per second: the membrane carrying the iron will vibrate as many times, and so many pulses of induced electricity will be *imposed* upon the constant current, which will each act upon the receiver, and cause so many vibrations of the armature upon it; and an ear held at P will hear the sound with the same pitch as that at the sending instrument. If two or more sound-waves act simultaneously upon the membrane, its motions must correspond with such combined motions; that is, its motions will be the resultant of all the sound-waves, and the corresponding pulsations in the current must reproduce at R the same effect. Now, when a person speaks in the tube, the membrane is thrown into vibrations more complex in structure than those just mentioned, differing only in number and intensity. The magnet will cause responses from even the minutest motion; and therefore an ear at R will hear what is said at the tube. This was the instrument exhibited at the Centennial Exposition at Philadelphia, and concerning which Sir William Thompson said on his return to England, "This is the greatest by far of all the marvels of the electric telegraph."

The popular impression has been, concerning the telephone, that the *sound* was in some way conveyed over the wire. It will be obvious to every one who may read this, that such is very far from being the case. The fact is, it is a beautiful example of the convertibility of forces from one form to another. There is first the initial vibratory mechanical motion of the air, which is imparted to the membrane carrying the iron. This motion is converted into electricity in the coil of wire surrounding the electro-magnet, and at the receiving-end is first effective as magnetism, which is again converted into vibratory motion of the iron armature, which motion is imparted to the air, and so becomes again a sound-wave in air like the original one.

This was the first speaking-telephone that was ever constructed, so far as the writer is aware, but it was not a practicable instrument. Many sounds were not reproduced at all, and, according to the report of the judges at the Philadelphia Exposition, one needed to shout himself hoarse in order that he might be heard at all.

THE AUTHOR'S TELEPHONE.

For several years past my regularly recurring duties have taken me over the various subjects treated of in this book, and each one has been extensively illustrated in an experimental way, and a considerable number of new pieces of apparatus and new experiments to exhibit their phenomena have been devised by me.

Among these, I would mention the following: —

1. Measurement of the elongation of a magnetized bar.
2. A magneto-electric telegraph.
3. An electro-magnetic instrument for demonstrating the rotation of the earth.
4. The permanent magnetism of the magnetic phantom.
5. The convertibility of sound into electricity.
6. The induction of a vibrating magnet upon an electric circuit.
7. The origination of electric waves in a circuit by a sounding magnet.
8. The discovery of the action of the air in a sounding organ-pipe.
9. Two or three methods for studying the vibrations of membranes.
10. Lissajous forks for enlarged projections of sound vibrations.

As soon, therefore, as I gave attention to the subject of telephony, I was able, with a few preliminary experiments, to determine the proper conditions for the transmission of speech in an electric circuit; and, without the slightest knowledge of the mechanism which Prof. Bell had used, I devised the following arrangement for a speaking-telephone.

My first speaking-telephone, Fig. 14, consisted of a magnet made out of half-inch round steel bent into a U form, having the poles about two inches apart. Over these were slipped two bobbins taken from an old telegraph register, and were already fitted to a half-inch core. These bobbins, two inches and a half long, were

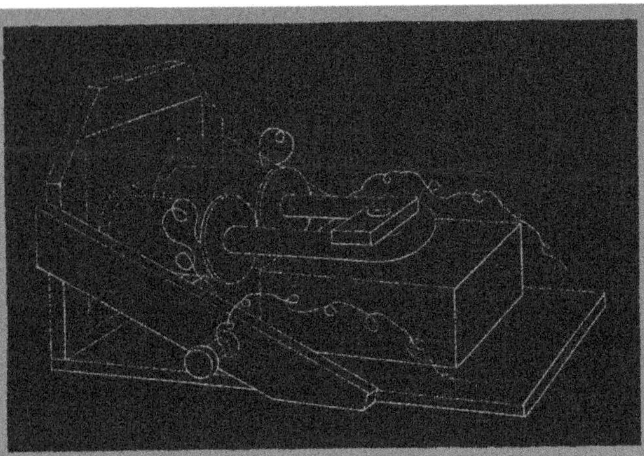

FIG. 14. — MY FIRST SPEAKING TELEPHONE.

wound with cotton-covered copper wire, No. 23,

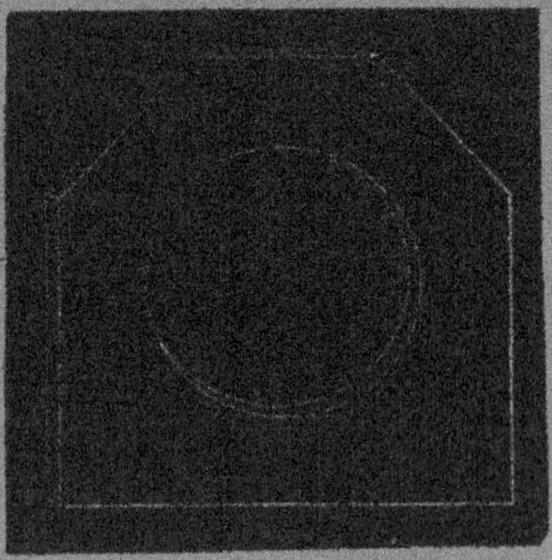

FIG. 14. — END VIEW.

each bobbin containing about 150 feet. This magnet, with the bobbins slipped upon its poles, was made fast to a post two or three inches high. The steel was made as strongly magnetic as was possible, and would hold up three or four times its own weight. In front of the poles, a sheet of thin steel, one-fiftieth of an inch thick, was made fast to an upright board having a hole cut through it three and a half inches in diameter (Fig. 14, end view); the plate was screwed tightly to this board, so as to cover the hole; and the middle of the hole was at the same height as the two poles of the magnet. The wires from the two bobbins were connected, as if to make an electro-magnet; while the two free terminals were to be connected with the line-wires. Of course there were two of these instruments, both alike; and talking and singing were reproduced with these.

A very great number of experiments have been made to determine the best conditions for each of the essential parts, — the size and strength of the magnet, the size of the bobbins, as to length and fineness of wire, the best thickness for the plate for absorbing the vibrations, &c.; and it is really surprising, how little is the difference between very wide limits. The following directions will enable any one to construct a speaking-telephone with which good results may be obtained. The specifications will be for only one instrument;

though of course two instruments made alike will be necessary for any purposes of speaking or other signals.

Procure three common horse-shoe magnets about six inches long, all of the same size; these

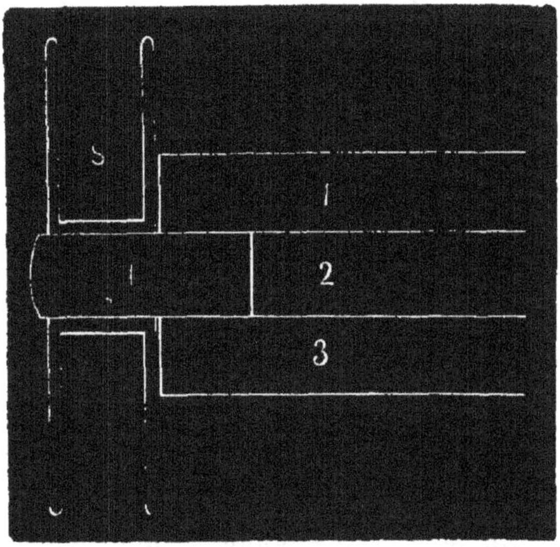

FIG. 15.

retail in the market at about a dollar apiece. They should be strong enough to hold up several times their own weight each. Next, have turned out of good hard wood, — such as maple or box-wood, — two spools not over half an inch long and an inch and a half broad, the sides cut square both inside and out, as shown at S, Fig. 15; a hole the third of an inch in diameter is to be made through the spool. Into this hole is to be fitted a short rod of soft iron, I, about an inch long, which should be a little rounded at the outer end. The bobbins may be wound with as much insulated copper wire as they will hold. The wire may be from the one-fortieth to the one-fiftieth of an inch in diameter, as is most convenient to obtain, the latter size being preferable. The resistance of such bobbins will probably be from two to three ohms each. The soft-iron core I must project backwards far enough to be clamped between the two outer magnets 1 and 3, while the inner one, 2, is drawn back. When the bobbins are in their places, and are clamped between the upper and lower magnets, they will stand as shown in Fig. 16, where the view is from above; the magnets being buttoned down to the block they rest on (see Fig. 17), which at the same time holds the soft-iron rods with the bobbins upon them. The wires on these coils must be connected in the same way they would be in order to make opposite poles of their outer ends, if a current of electricity were to be sent through the coils. An upright board B (Fig. 17) six or seven inches square, having a round hole four inches in diameter cut out from the middle of it, must be fixed near the end of the base-board ; and over this hole is to be screwed *tightly* a piece of thin

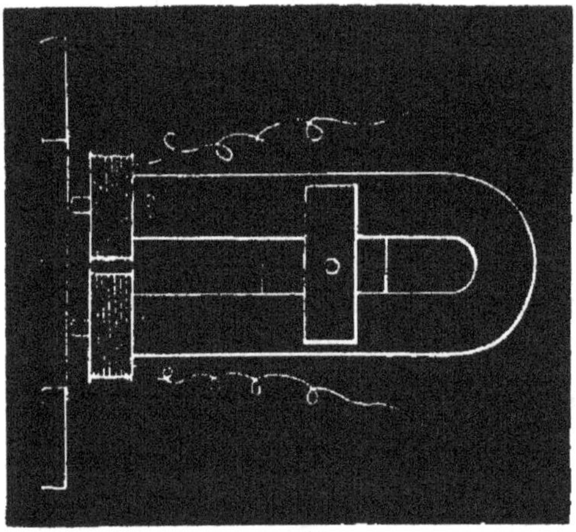

FIG. 16.

sheet iron or steel ; it may be from the one-twentieth to the one-fiftieth of an inch in thickness. It does not seem to make much difference about the thickness of this plate. I have generally got the best results from a plate one-fiftieth of an inch thick. The upright board carrying this plate must be very rigid, otherwise the plate will be kept tight to the magnets all the time ; and one of the conditions of success in working is, that this plate shall be as close as possible to the magnet-ends, but not to touch : therefore fix the board tight, and adjust the magnets by means of the button shown on top of them in the perspective figure.

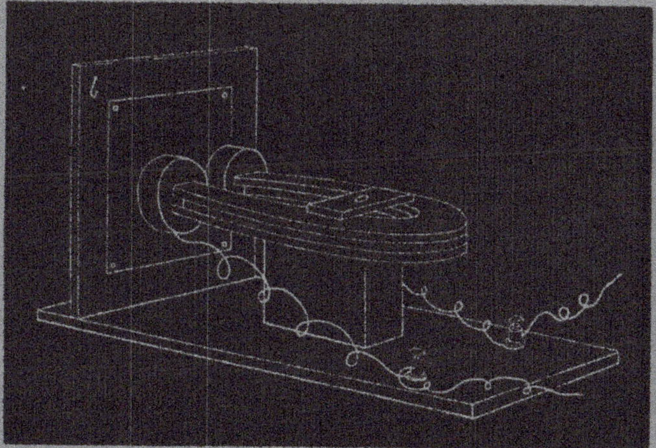

FIG. 17.

The sounds to be transmitted, of whatever sort they may be, are to be made on the side P, Fig. 16 ; and likewise, when the instrument is used as a receiver, the ear is to be applied at the same place. A tube about two inches in diameter may be made fast to the front of the board, in a line with the centre of the plate; this will aid somewhat in hearing. When two or three persons are to sing, it will be best to have each one supplied with a tube to sing through; one end of the tube to be placed close to the front of the plate. The sound of musical instruments, such as the flute and the cornet, will be reproduced much louder, if the front of such instrument be allowed to rest upon the rim of the hole in the board, just in front of the plate.

It is noticeable that low talking can be heard more distinctly than when a great effort is made; but the sounds though distinct are not strong at any time, and other sounds seriously interfere with hearing. It is probable that some way will hereafter be devised for increasing the usefulness of the invention by increasing the volume of sound. On account of the weakness of the sound it becomes necessary to provide a call to attract the attention of one in the room. This may be accomplished by having a small electric bell worked by a one or two cell battery. Another way which I have found to be quite as efficient is to have a rod of iron or steel about a foot long, and half an inch in diameter, bent into a U form. When this is held by the bend, and struck upon the floor or with a stick, it vibrates powerfully; and if one of its prongs be permitted to strike against the plate P, Fig. 16, the sound will be reproduced loud enough to hear over a large room. I have never failed to call with this when any one was in the same room with the telephone.

Wherever a telephone circuit has been made upon telegraph poles having other wires upon them, the inductive actions of the currents upon the other wires has been found to seriously interfere with the action of the telephones, inasmuch as the latter reproduce every other message. One skilled in reading by sound in the ordinary way can read through the telephone what message is travelling in a neighboring wire.. Messages may be thus read upon wires as far distant as ten feet from the telephone circuit. It therefore seems to be essential that each telephone circuit should be isolated from every other one, else there can be no secrecy in messages.

A very interesting effect was noticed one night

when there was a bright aurora display. There was a continuous current through the wires, accompanied with sounds which increased in intensity as the bright streamers passed by. This will probably lead to some important results in science.

In all probability the telephone is as much in its infancy as was ordinary telegraphy in 1840. Since that time the sciences of electricity and magnetism have had the most of their growth, and telegraphy has kept pace with the advancing knowledge until its commercial importance is second to no other agency. Very many important principles that are invaluable in telegraphy to-day were wholly unknown in 1840; but it may here be noted that in the telephone, as it now is, there is not a single principle that was not well enough known in 1840. This will be apparent to one who follows out the phenomena from the sender to the receiver. First, the sound in air causing a corresponding movement in a solid body, iron. This iron, acting inductively upon a magnet, originates magneto-electric currents in a wire helix about it; and these travel to another helix, and, re-acting upon the magnet in it, have electro-magnetic effects, and increase and decrease the strength of the magnet; and this variable magnetism affects the plate of iron in front of that magnet, and makes it to vibrate in a corresponding manner, and thus to restore to the air in one place the vibrations absorbed from the air in another place. To some it may seem strange that a simple thing as the telephone is, involving nothing but principles familiar enough to every one interested in physical science, should have waited nearly forty years to be invented. The reason is probably this: Men of science, as a rule, do not feel called upon to apply the principles which they may discover. They are content to be *discovering*, not *inventing*. Now, the schools of the country ought to make the youth quite familiar with the general principles of physical science, that the inventive ones — and there are many such — may apply them intelligently. Mechanism is all that stands between us and aërial navigation; all that is necessary to reproduce human speech in writing; and all that is needed to realize completely the prophetic picture of the "Graphic," of the orator who shall at the same instant address an audience in every city in the world.

Made in the USA
Monee, IL
04 November 2026

67385021R00075